T0325230

Measuring Behaviour

Measuring Behaviour is the established go-to text for anyone interested in the scientific methods for studying the behaviour of animals or humans. It is widely used by students, teachers and researchers in a variety of fields, including biology, psychology, the social sciences and medicine.

This new fourth edition has been completely rewritten and reorganised to reflect major developments in how behavioural studies are conducted. It includes new sections on the replication crisis, covering Open Science initiatives such as preregistration, as well as fully up-to-date information on the use of remote sensors, big data and artificial intelligence in capturing and analysing behaviour. The sections on the analysis and interpretation of data have been rewritten to align with current practices, with clear advice on avoiding common pitfalls.

Although fully revised and revamped, this new edition retains the simplicity, clarity and conciseness that have made *Measuring Behaviour* a classic since the first edition appeared more than 30 years ago.

Professor Melissa Bateson studied at the University of Oxford, graduating with an MA in zoology with biological anthropology and a DPhil in animal behaviour. She held a Wellcome Trust fellowship in the Department of Zoology at Oxford and the Department of Psychology at Duke University, USA, followed by a Royal Society University Research Fellowship in the Department of Psychology at Newcastle University, UK. She is currently Professor of Ethology at Newcastle where she teaches behaviour at undergraduate and postgraduate levels and runs an active research programme. Melissa is the daughter of the late Patrick Bateson and replaces him as co-author on this fourth edition.

Dr Paul Martin studied biology at the University of Cambridge, graduating in natural sciences and with a PhD in behavioural biology, and was a Harkness Fellow in the Department of Psychiatry and Behavioral Sciences at Stanford University. He subsequently lectured and researched in behavioural biology at the University of Cambridge and was a Fellow of Wolfson College. He co-authored the first three editions of *Measuring Behaviour* with Patrick Bateson, and is the author or co-author of several other books including *Design for a Life* (2000), *Counting Sheep* (2002) and *Play, Playfulness, Creativity and Innovation* (2013).

Measuring Behaviour

An Introductory Guide

Melissa Bateson
Paul Martin

CAMBRIDGE
UNIVERSITY PRESS

CAMBRIDGE
UNIVERSITY PRESS

University Printing House, Cambridge CB2 8BS, United Kingdom

One Liberty Plaza, 20th Floor, New York, NY 10006, USA

477 Williamstown Road, Port Melbourne, VIC 3207, Australia

314–321, 3rd Floor, Plot 3, Splendor Forum, Jasola District Centre, New Delhi – 110025, India

79 Anson Road, #06–04/06, Singapore 079906

Cambridge University Press is part of the University of Cambridge.

It furthers the University's mission by disseminating knowledge in the pursuit of education, learning, and research at the highest international levels of excellence.

www.cambridge.org
Information on this title: www.cambridge.org/9781108478311
DOI: 10.1017/9781108776462

© Cambridge University Press 2021

First published 2021

A catalogue record for this publication is available from the British Library.

ISBN 978-1-108-47831-1 Hardback
ISBN 978-1-108-74572-7 Paperback

This book is dedicated to the memory of
Professor Sir Patrick Bateson FRS (1938–2017)

Professor of Ethology at the University of Cambridge, Vice President of the Royal Society, Provost of King's College Cambridge, Director of the University of Cambridge Sub-Department of Animal Behaviour, President of the Association for the Study of Animal Behaviour, President of the Zoological Society of London and co-author with Paul Martin of the first three editions of *Measuring Behaviour*, published in 1986, 1993 and 2007. Patrick Bateson was a distinguished scientist whose work advanced the understanding of the biological origins of behaviour. He will also be remembered as a man of immense warmth and kindness, whose success as a leader, teacher and administrator owed much to his collaborative spirit, generosity and good humour.

Contents

Acknowledgements

The authors would like to thank the following individuals who helped in various ways with the writing and publication of this new edition of *Measuring Behaviour*: Dusha Bateson, Olivia Boult, Evie Button, Damien Farine, Jordan Giddings, William Hoppitt, Charles Howell, Jane Hoyle, Megan Keirnan, Kate Lessells, Georgia Mason, David Massey, Daniel Nettle, Liz Paul, Colline Poirier, Mark Prescott and Claire Witham.

1

Introduction

1.1 Scope and Aims

As its title suggests, this book is about measuring behaviour. By behaviour, we mean the actions and reactions of whole organisms or groups of organisms. Examples of behaviour include the mating displays of bower birds in the jungle, the foraging patterns of laboratory mice and the productivity of humans working in an office. The organisms whose behaviour is measured are usually multicellular animals such as insects, birds and mammals, including our own species, *Homo sapiens*. However, the possession of a brain is not a prerequisite for demonstrating measurable behaviour: unicellular organisms such as bacteria and protozoa, and even plants, are increasingly targets for behavioural research [1]. The 'organisms' need not even be biological: they might be artificial intelligence (AI) systems embodied in robots or autonomous vehicles, or virtual agents behaving in the virtual environment of a computer [2]. This book is intended to be relevant to anyone interested in measuring behaviour in any of these diverse entities, which we refer to as 'subjects'.

Measuring behaviour accurately and reliably requires various problems to be solved. General principles underlying the measurement and analysis of behaviour apply whether the subjects are microbes, animals, people or machines. Many of the same problems and principles apply whether the behaviour being studied is occurring in the natural environment, the laboratory or a computer. Our aim is to describe these problems and principles and offer practical advice to anyone who wants to understand how behaviour is measured.

The main focus of this book is on the direct observation of behaviour. By direct observation, we mean the description and analysis of what subjects actually do in a specific situation. Direct observation does not imply that the behaviour must be directly observed in real time but rather that the variables recorded must relate directly to the actual performance of the behaviour in question. In the past, behaviour was generally measured by human observers using low-tech methods such as check sheets, voice recorders or computer event recorders, and such methods are often still appropriate. Increasingly, however, researchers are using automated methods such as machine vision, data loggers attached to subjects and, for human subjects, smartphones.

Some behaviour patterns may leave semi-permanent evidence that can be measured as a proxy for the behaviour of interest. For example, the structure of a bird's nest provides a record of its construction decisions; bald patches of skin on a mouse can be evidence of over-grooming by cage mates; and a photograph of a meal may provide a reasonable proxy for human diet choice.

In non-human animals, the direct observation of behaviour or its immediate proxies is the only option available for measuring behaviour. For humans, however, surveys, interviews and questionnaires are commonly used to measure how subjects believe they behaved or might behave in given situations. While such methods involve the performance of behaviour – the subject has to make verbal or written responses – the behaviour that is directly observed (e.g. ticking a box on a questionnaire) is distinct from the actual performance of the behaviour of interest (i.e. the behaviour described in the questionnaire) [3]. Measurements based on self-reported memories, or beliefs about probable behaviour, may be justified on the grounds that they are less time consuming than direct observation and may correlate reasonably well with actual behaviour. Nonetheless, retrospective self-report is often very different from what people actually did, owing to a combination of inaccurate memory and biases in what people believe about how they behave, or what they choose to convey to researchers. For example, total calorie consumption is systematically underreported by obese human subjects [4]. Alcohol consumption is also systematically underreported, with retrospective reports becoming more inaccurate the more subjects drank on the day they were asked to recall [5]. Retrospective self-report measures therefore provide an indirect and often inaccurate means of measuring behaviour, though they are sometimes the only feasible means available.

Previous editions of this book were focused on measuring behaviour in non-human species. However, we wanted this new edition to be equally relevant to the direct observation of behaviour in humans. Modern technologies offer novel approaches to the direct observation of human behaviour that circumvent traditional obstacles. For example, wearable fitness-tracking devices can measure actual physical activity, and smartphones can be used to ask subjects what they are doing in real time. Although the latter method relies on self-report, it is less susceptible to inaccuracies resulting from poor memory and also to biases resulting from what is socially or medically acceptable. For instance, admitting to drinking three units of alcohol that evening is still well within an advisory limit, whereas 21 units that week puts the subject over the current UK advisory limit and may be information the subject would choose not to share.

Measurement means recording phenomena in a systematic, reproducible way and expressing the results in numerical form. Measuring behaviour therefore implies a **quantitative** approach. Social scientists make a distinction between quantitative and **qualitative** research methods. To explain the difference, let us consider how quantitative and qualitative researchers might explore whether a group of chimpanzees is unusually aggressive. Both types of researcher would spend time watching the animals. The quantitative researcher would then define a set of aggressive behaviours (e.g. chases, threats, bites) and count the occurrence of these behaviours observed in a specified time period. They would use statistical methods to summarise the number of aggressive behaviours seen in the group and compare the results to comparable data collected from other groups to test the hypothesis of a difference in aggression. The qualitative researcher would describe verbally what they observed without the constraint of sticking to predefined categories. They would identify themes in aggressive behaviour (e.g. male–male aggression, inter-sexual conflict, abuse of infants) and use specific examples to illustrate them (e.g. Bobby bit Kathy on the arm when she attempted to grab his food). Based on this data, they would reach a subjective impression of the types of aggression displayed and the aggressiveness of the group. Thus, the distinction between qualitative and quantitative research comes down to the methods used to represent, summarise and analyse the data collected.

Qualitative approaches give a richer impression of the behaviour, but quantitative approaches make it clearer how conclusions follow from the data because they are more transparent and reproducible. Furthermore, quantitative data allows statistical comparisons to be made between groups and hypotheses to be formally tested. Worthwhile research on behaviour will require that at least some aspects of behaviour are measured, and we would advocate a quantitative approach wherever possible. That said, all quantitative measurements are ultimately based on subjective judgements about what to ask and what to record. The art of measuring behaviour is learning to ask the right questions and to choose good behavioural metrics.

1.2 Why Measure Behaviour?

Behaviour is centrally important in many areas of biology, psychology, medicine and the social sciences. Within the biological sciences, behaviour sits at the nexus between neuroscience, cognitive science and ecology. Behaviour is the major output of the brain and is therefore central to unravelling the cognitive and neural mechanisms underlying it. Behaviour is also a major means by which individual organisms adapt to and affect their physical and biological environments, making it a major factor in ecology. Human behaviour is, of course, central to psychology and most fields of social science. Anthropology, communication, education, economics, human geography, law, linguistics, political science, public health, security and sociology all involve the description and understanding of human behaviour and its consequences. Many of the biggest problems facing the human species, including climate change, infectious disease pandemics, obesity, mental health, geopolitical conflict and terrorism, are reflections of human behaviour. The vital importance of measuring how people behave was highlighted by the COVID-19 pandemic. Accurate and reliable descriptions of the behaviour underlying such problems are critical to understanding them and implementing the changes necessary to address them. Given the central importance of behaviour in so many disciplines, researchers must understand how to measure it accurately and reliably, while policy makers must understand how to distinguish between good and bad measurements.

1.3 What is So Special About Measuring Behaviour?

The answer to this question lies in the characteristics of behaviour that make its measurement challenging. Behaviour has a temporal component – it unfolds over time. Therefore, it is rarely possible to measure behaviour in a meaningful way by looking at a single snapshot in time; the methods used must accommodate the temporal component. Occasionally, there may be a simple proxy for a behaviour that *can* be measured at a single time point. For example, a single blood test for the level of cotinine (the main metabolite of nicotine) can reveal how much a person smokes on average. But it cannot reveal potentially important details about the pattern of smoking behaviour, such as the time of day at which the behaviour occurs or its social context.

Many behaviour types are not discrete, making it hard to determine exactly when they begin and end. Take play behaviour, for example. It can be relatively easy to recognise that play has occurred within a given time window but difficult to define its exact duration.

Behaviour patterns are often extremely complicated, making them difficult to represent with a single number or metric. Counting the stereotyped distress calls of a domestic chick is straightforward, whereas describing and quantifying the complex song of a nightingale in a meaningful way is considerably harder.

Behaviour often changes in response to stimuli from the environment. Moreover, different species have very different sensory systems, which means that stimuli that are undetectable to humans may be highly salient to other species. For instance, birds and bees can see in the ultraviolet (UV) part of the spectrum, while mice and rats can hear ultrasound. Even humans may respond to stimuli of which they are not consciously aware. For example, people's reactions to a movie may be heightened in the cinema by exposure to volatile organic compounds released by other viewers in response to humour or suspense [6].

A further important characteristic of behaviour is that it is highly variable. This variability occurs at different levels: there is variation in behaviour between individuals of the same species, depending on the individual's genes, sex, developmental history and so on; there is variation within individuals over time resulting from the effects of experience, maturation, developmental plasticity and senescence, among other things; and there is variation within individuals according to their current context. Behaviour varies according to physical variables, such as time of day and ambient temperature, and biological variables, such as hormonal state and social context. While some sources of variability will be obvious factors that can be accounted for when designing a study, others may be less apparent.

For example, the way in which a researcher handles a mouse when placing it in the test apparatus is known to affect the animal's anxiety levels even more than conventional laboratory manipulations of stress [7]. Similarly, the sex of the human experimenter present in the room is found to influence the behaviour of laboratory rodents in standardised tests of anxiety, probably via the volatile pheromones released by men and women [8]. These and other characteristics of behaviour make its measurement challenging. An awareness of the potential pitfalls is critical in designing and conducting a good study.

1.4 Steps Involved in Studying Behaviour

Those attempting to make systematic measurements of behaviour for the first time are often daunted by the apparent difficulty of the job facing them. How will they ever notice, let alone record accurately and systematically, everything that is happening? The truth is that measuring behaviour *is* a skill but not one that is especially difficult to master, given some basic knowledge and an awareness of the possible pitfalls.

Studying behaviour involves a number of inter-related processes that can be broken down into a series of steps. Although the steps listed below will apply in the majority of behavioural studies, the order in which they are taken will vary. Moreover, some steps may need to be repeated multiple times in the light of results obtained from preliminary observations or pilot studies. Box 1.1 gives an illustrative example of how the steps apply in an actual study. The steps involved in studying behaviour can be summarised as follows:

1. **Ask a question.** All scientific studies should start with a clear question.
2. **Formulate hypotheses.** Hypotheses are provisional and testable explanations for observed phenomena. A hypothesis can be thought of as a possible answer to the question. At least two alternative hypotheses should be formulated. Good hypotheses should give rise to testable predictions.
3. **Make predictions.** A clear hypothesis should, by a process of straightforward reasoning, give rise to one or more specific predictions that can be tested empirically. Developing specific predictions goes hand in hand with identifying which behavioural variables to measure.

Box 1.1 Example of the steps involved in a behavioural study exploring the relationship between self-reported food insecurity and measured snack food consumption in adult British women [9]

1. **Question.** Why is current food insecurity – defined as the limited or uncertain ability to acquire nutritionally adequate food – associated with obesity in women in affluent Western societies?
2. **Hypotheses.** Main hypothesis tested: food-insecure women are motivated to consume more calories than they require when the opportunity is available. Null hypothesis: the calorie consumption of food-insecure and food-secure women does not differ.
3. **Prediction.** When presented with a controlled opportunity to consume snack foods in the laboratory, food-insecure women (as defined by their responses to a standard questionnaire) will consume more calories than food-secure women.
4. **Behavioural metric.** The total calories in three snack foods (chocolate, crisps and popcorn) consumed by a participant in a fixed time period.
5. **Recording method.** Participants were given an opportunity to snack via the guise of a mock taste test in which they were asked to evaluate aspects of the palatability of three different snack foods. Plates of the three foods presented to the participants for evaluation were weighed on an electronic balance by the researchers before and after the test.
6. **Study design.** An observational study using naturally occurring variation in the food insecurity of an opportunity sample of participants recruited at Newcastle University. Recruitment of participants was time limited, but a minimum target number required was set prior to commencing data collection.
7. **Ethics.** The study was approved by the Faculty of Medical Sciences ethics committee of Newcastle University. All participants provided written informed consent and were debriefed after the study.
8. **Pilot study.** No formal pilot study was conducted in this case. Prior to collecting data, all the researchers involved in data collection rehearsed the protocol to ensure standardised procedures.
9. **Preregistration.** This study was not preregistered. It was therefore constrained to be published as an exploratory study.
10. **Data collection.** An agreed protocol was followed throughout, including a standard script for instructing participants. Researchers measured snack food consumption blind to the food-insecurity status of the participants. Data was collected from 84 women.

11. **Data analysis.** The hypothesis was tested using a general linear model, with total calories consumed as the outcome variable and food insecurity as a continuous predictor variable. As predicted, higher adult food insecurity was found to be associated with greater total calorie consumption in the mock taste test. Additional exploratory data analysis showed that greater self-reported childhood food insecurity reduced the positive effect of adult food insecurity on calorie consumption.

12. **Communication.** The study was initially written up independently by three undergraduate psychology students as a requirement for their degree course. Subsequently, their supervisors prepared a version of the study for publication in a peer-reviewed academic journal (*Appetite*). The raw data and an accompanying script for the data analysis were made publicly available via the Zenodo repository.

4. **Identify and define behavioural metrics.** Behavioural metrics must be defined clearly and unambiguously before starting to collect data. Metrics may need to be redefined or replaced in the light of preliminary observations and pilot measurements.

5. **Choose an appropriate recording method.** How will the metrics identified in step 4 actually be recorded to ensure accurate and reliable data? The method chosen will depend on the nature of the behaviour to be recorded and the technology available. In practice, steps 4 and 5 are inter-related, as the recording method often influences the choice and definition of metrics.

6. **Design the study.** This involves making choices about the type of study to be conducted (observing natural variation versus experimental manipulation), the selection of subjects (e.g. mice or humans), the allocation of subjects to experimental groups (sampling and random-isation), how many subjects are required and how data collection will be structured (cross-sectional versus longitudinal). Good design makes the difference between a study that allows hypotheses to be tested and one that is uninterpretable. It can also affect costs in terms of time, money and subject welfare, with implications for step 7.

7. **Ensure that the research is legally and ethically compliant.** Most behavioural research will require formal ethical permission to be obtained from

the relevant national or institutional regulatory bodies before it can proceed. A successful application for ethical approval will usually require that steps 1–6 have already been completed.

8. **Conduct pilot studies.** Jumping straight in and collecting 'hard data' from the very beginning is rarely the best way to proceed. Preliminary observations can be useful for formulating and sharpening hypotheses and refining behavioural metrics. Pilot studies allow researchers to practise experimental manipulations and recording methods and assess the reliability and validity of measurements.

9. **Preregister the study.** To reduce dubious research practices (see Chapter 2), it is good practice to preregister the hypotheses, predictions and methods for a study in a publicly accessible place before starting to collect data.

10. **Collect the data.** The same measurement procedures should be used throughout. If possible, data should be collected 'blind', so that measurements are not unconsciously selected or adjusted to fit the hypotheses. Data collection should stop when a predetermined threshold (identified in step 6) has been reached.

11. **Analyse and interpret the data.** The data collected should be thoroughly visualised and understood through the use of graphs and descriptive statistics before embarking on statistical hypothesis testing. Confirmatory data analysis is used to test preregistered hypotheses. Exploratory data analysis is used to obtain the maximum amount of information from the data and discover unexpected results that may generate new questions.

12. **Write up and communicate the research.** Research is wasted if it is not properly communicated. Where research has used sentient subjects, it is ethically dubious not to make the data and findings available because the subjects have given their time and in some cases their lives in the cause of research. Many research funders require data to be made publicly available.

In the following chapters we add flesh to the skeleton outlined above. Chapter 2 considers the quality of behavioural research and describes the practices necessary to ensure replicable findings. Chapter 3 distinguishes four logically distinct questions that can be asked about any behaviour and considers how a question is turned into testable hypotheses and predictions.

Chapter 4 discusses alternative study designs and the problem of how much data to collect. Chapter 5 describes the legal and ethical issues raised by behavioural research and the requirements before a study can proceed. Chapter 6 describes the various metrics available for quantifying behaviour and Chapter 7 the alternative rules for sampling and recording behaviour. Chapter 8 describes the technologies that assist data capture and processing. Chapter 9 discusses the problems and opportunities raised by the fact that subjects often occur in groups. Chapter 10 describes methods for ensuring the validity and reliability of behavioural measurements. Chapter 11 introduces how behavioural data is analysed. Finally, Chapter 12 considers pitfalls in the interpretation of findings and describes how research is communicated.

1.5 Summary

- Behaviour is the actions and reactions of an organism or group of organisms. Living organisms, robots and virtual agents all exhibit measurable forms of behaviour.
- Measuring behaviour involves assigning numbers to direct observations of behaviour using specified rules.
- Direct observation means collecting data that relates directly to the performance of the behaviour pattern in question.
- Measuring behaviour accurately and reliably is important because behaviour is central to answering many questions in the biological and social sciences.
- Measuring behaviour is challenging because behaviour has a temporal component, does not always occur in discrete bouts, is generally complicated, can be influenced by stimuli undetectable to humans, and varies both within and between individuals.
- Studying behaviour can be broken down into a series of steps that starts with asking a question and ends with communicating findings.

2
Science and Truth

2.1 The Replication Crisis

Science is about discovering truth. The ultimate aim of any behavioural study is to discover true facts about the behaviour under investigation and produce evidence that has a bearing on a **scientific theory** (see Box 2.1 for definitions). Incorrect or faulty theories present no great threat to science in the long run because they will eventually be found to be at odds with reality: observational studies or experiments will ultimately produce empirical results that are incompatible with the theory, forcing the theory to be modified or even abandoned. Faulty empirical findings are more insidious because they may be widely accepted as true unless explicit attempts are made to reproduce them.

When empirical results bear crucially on an important theory, it is standard scientific practice for measurements to be replicated. Empirical findings should be checked and double-checked if an important theory stands or falls by them. In the behavioural sciences, however, the immense diversity, variability and complexity of behaviour, together with the relative dearth of quantitative theories, have historically reduced the pressure to replicate. As a consequence, incorrect and misleading empirical results have, until recently, gone unchecked and unchallenged. Over the past decade, the magnitude of this problem has started to emerge and we are currently facing what has been called a **replication crisis** [10, 11]. Recent replication studies in psychology and the social sciences have revealed that less than half of the findings published in the top journals could be successfully replicated [12–14]. Even when findings were replicated, the **effect sizes** were often much smaller than those reported in the original studies; that is, the differences between groups were smaller, or the correlations between variables were lower, than those originally reported.

This replication crisis has had a major impact within the behavioural sciences [15]. It has affected how the quality of existing scientific evidence is assessed and the methodology for doing new research. Perhaps the most

Box 2.1 Definitions of key terms

Scientific theory. A currently accepted explanation or framework for understanding a set of empirical observations. Theories are developed and tested following an accumulation of relevant empirical evidence.

Hypothesis. A clear statement describing a plausible candidate explanation for empirical observations. The hypothesis under test in a study is usually contrasted with a **null hypothesis**, the hypothesis of no effect. Hypotheses follow from theories and lead to testable predictions.

Effect size. A metric that quantifies the size of the difference in the means of two groups or the strength of a correlation between two variables. Larger effects are characterised by bigger differences in means or stronger correlations between variables. **Standardised effect sizes** (such as Cohen's d) express these latter differences in units of standard deviations, and hence allow estimates from different studies to be compared directly.

Sample size. A measure of the number of replicates in a study. In many simple designs, in which each subject contributes one measurement to the dataset, sample size is equal to the number of subjects in the study. Sample size is commonly referred to as n.

Random variation. The variation between measurements that cannot be accounted for by the variables manipulated or controlled in the study and which is therefore considered to be due to other (random) factors. Measurement error is one common source of random variation. Random variation is sometimes referred to as **noise**.

P-value. A number derived from a statistical test that estimates the probability of obtaining an effect at least as large as the observed effect if the true effect is in fact zero (i.e. the null hypothesis is true). The conventional **criterion for statistical significance** is $p < 0.05$ and is referred to as α (the **false-positive** or **type I error** rate; see Table 2.1).

Statistical significance. A statistically significant result is one that is very unlikely if the null hypothesis is true, with 'very unlikely' conventionally being defined as having a probability of less than one in 20 (i.e. $p < 0.05$). Statistical significance is not the same as biological, clinical or social significance. For example, there may be a reliable but very small effect of a drug on some biomarker in patients with a particular disease, which is statistically significant when compared with the effect of a placebo, even though the drug effect is insufficient to deliver any clinical benefits in terms of symptom reduction.

Statistical power. The probability that a study will detect an effect when there is a true effect to be detected. This is the same as the probability of not making a **false-negative** or **type II error** (Table 2.1).

> The false-negative error rate is sometimes referred to as β and power is thus equal to $1 - \beta$. The power of a study is affected by the effect size to be detected, the amount of random variation and the sample size (number of replicates). A larger sample size always results in higher power.

Table 2.1 Four possible outcomes from a scientific study

		Objective truth	
		Hypothesis false	Hypothesis true
Study findings	Hypothesis rejected	True negative	False negative[a]
	Hypothesis confirmed	False positive[a]	True positive

[a] False-positive and false-negative results are also referred to as type I and type II errors, respectively.

damaging outcome of the replication crisis has been its contribution to the popular misconception that behavioural science is 'soft', in contrast to the 'hard' science done by physicists. This is misleading because all science, whether physics or psychology, can be done well or badly. Worryingly, as we will see below, bad science is at least partly to blame for the replication crisis. But just because some behavioural science has not been done well does not, of course, mean that all of it is bad. The scientific method is still the best way of discovering the truth, and behavioural scientists need the skills to recognise and conduct good science.

2.2 What Does Replication Mean?

Replication may seem rather dull – why waste time and money replicating results that are already published, rather than forging ahead with something new? Nonetheless, the replication crisis has served to emphasise that replication is an essential cornerstone of good science. A key result should be replicated before attempting to build on it, while a failed replication attempt may well save resources in the longer term by avoiding building a research programme on shaky foundations.

There are several complementary approaches to replication, all of which can contribute to establishing scientific truths. The first simply is to check whether the same results and conclusions can be reached using the *exact same* data. This is sometimes referred to as **reproducibility**, to distinguish it from other types of replication. While it might seem inevitable that an independent reanalysis of the same data would produce the same result, in reality this is far from certain. The analysis and interpretation of a dataset involves a series of decisions over which data to include, which statistical tests to perform and how to interpret the results. This flexibility is sometimes referred to as **researcher degrees of freedom**. Different researchers may make different decisions, potentially leading to different conclusions. The growing trend for researchers to make available their raw data as part of the publication process makes an analysis of reproducibility a sensible place to start any replication attempt.

True replication investigates whether the same result is obtained if a study is repeated using the same procedures to yield new data. One approach is to duplicate a particular study using the same species or population of subjects and precisely the same procedures, conditions and metrics as the original study. This approach, which is referred to as **literal replication**, is important for checking that the original results were not a statistical fluke. Until relatively recently, literal replication studies were rarely conducted in the behavioural sciences because researchers were not incentivised to do them.

One problem with literal replication in behavioural research is that no two studies are truly identical: even the same sample of subjects measured under the same conditions may vary with factors such as time or experience. Thus, exact replication under truly identical conditions is rarely feasible in practice. A second problem with literal replication is that it may give rise to unjustified confidence about the generality of findings that in fact rely on one aspect of the methodology. For example, many female birds of a given species may respond in the same way to one recording of male song. However, they might have responded differently to another recording that is subtly different [16]. If the aim is to make an inference about how females respond to typical male song, it is necessary to test female responses to a random sample of male songs representative of the population of male songs, not just a single exemplar. An otherwise well-planned experiment may fail to generalise because only a very restricted conclusion may be drawn from the results relating to the single stimulus actually tested.

A different approach to replication, known as **constructive replication** or **triangulation**, involves attempting to produce convergent results that lend support to the initial study's findings but using different procedures, different metrics, different experimental conditions or even a different species. No explicit attempt is made to duplicate the original study exactly; the emphasis instead is on whether the original conclusions are supported by complementary sources of evidence [17]. The strength of this approach is that if results are found to agree across different methodologies, they are less likely to be artefacts of one particular methodology. For example, the biological theory of kin selection – whereby natural selection favours genetic variants that predispose individuals to preferentially direct help towards related individuals – is supported by diverse converging evidence from a range of species.

2.3 Why Does Replication Sometimes Fail?

There are several reasons why a replication attempt may fail to produce the same result as the original study [18, 19].

2.3.1 Variability in Important Moderators

The first possibility to consider is that the replication was not a literal replication: some unconsidered variable differed between the original study and the replication attempt, giving rise to the difference in findings. Such variables are called **moderators**.

Behaviour is extremely variable. One source of variability is that behaviour is moderated by many factors, including species, breed, current physical environment, the presence of human observers and even the sex of human observers. Moreover, individuals, populations or strains of the same species may have different **reaction norms**, responding differently to variation in current environmental conditions. It is therefore entirely plausible that some failures of replication can be explained by unidentified moderators of behaviour. For example, a study on two mouse strains found varying effects of genetic strain on behaviour depending on the time of day at which the mice were tested, suggesting that both strain and time of day are important moderators of laboratory mouse behaviour [20].

The potential influence of moderators implies, somewhat counterintuitively, that over-standardisation of laboratory protocols could actually reduce the replicability of results – a phenomenon referred to as the **standardisation fallacy** [21]. If one laboratory always tests their mice in the morning and another always tests in the afternoon, their results may be different. However, if both laboratories spread their testing over the day, they would obtain the same mean result, even though their results would be more variable. There is often a trade-off between standardisation, to reduce variability in results, and what has been called 'systematic heterogenisation', to increase the generality of findings. The appropriate level of standardisation for a particular study will depend critically on the research question (see Chapter 3).

2.3.2 Low Statistical Power

Why might replication fail, even when there are no differences in relevant moderators and the original methods are followed exactly? The answer lies in understanding how the findings from a single scientific study might relate to the objective truth (Table 2.1). Consider, for example, a study published in a reputable journal reporting that peacocks with more eyespots in their trains attract more peahens for mating. Here, the **hypothesis** under test was that peahens prefer peacocks with more elaborate trains. This hypothesis could either be true or false. The study finding was therefore either a true-positive or a false-positive result. If a subsequent replication study fails to reproduce the original finding, this must be either a true-negative or false-negative result.

Good scientific methodology aims to reduce the probability of both false-positive and false-negative errors. The rate of false-positive errors is controlled by choosing a conservative criterion for accepting a positive result as being **statistically significant**. This criterion is conventionally set at a **p-value** of less than 0.05, meaning that the false-positive error rate is set to be less than one in 20. The false-negative error rate is controlled by the **sample size** of the study: a larger sample size will result in a lower false-negative error rate.

Even with the best methodology, some errors will still occur as a result of chance alone. Peahens may indeed on average prefer males with more eyespots, but the effect could be small. There will be some **random variation** in peahen choice: females may sometimes choose males with fewer eyespots, perhaps because their ability to correctly estimate, remember or compare numbers of eyespots is imprecise. Behavioural data is also often subject to random variation because of imprecision arising from the measurement techniques (see Chapter 10). A weak effect on behaviour that has a lot of random variation will be hard to detect reliably, especially with a limited sample size. Such studies are described as underpowered because they lack sufficient **statistical power** to reliably reject the null hypothesis. For a variety of reasons, including constraints on money, time or availability of subjects, many behavioural studies turn out to have been underpowered and therefore yielded false-negative results (Box 2.2).

Even when a small study discovers a true-positive result, the estimated effect size is likely to be exaggerated – a phenomenon known as the **winner's curse**. This inflation in estimated effect size is worst for small, underpowered studies because such studies can only detect effects that happen to be large. If you toss a coin six times, only the most extreme result of getting all heads or all tails would allow you to conclude that the coin is biased (and you would estimate the size of the bias as 1.00), whereas if you toss the coin 100 times, the criterion for significance (at $p < 0.05$) is only 59 or more heads or tails (and you would estimate the size of the bias as 0.59 based on 59 heads). Thus, the lucky scientist who obtains a true-positive finding in an underpowered study is cursed by also obtaining an inflated estimate of the true effect size.

The winner's curse means that replication studies based on small sample sizes similar to the original study will often be underpowered and fail to reproduce the original findings, even if these were true. In recent replication studies, researchers have set out to reduce the probability of obtaining false-negative results by replicating the original experiments multiple times and by using much larger sample sizes than the original studies [14]. As expected from the winner's curse, when results were replicated in this way, the measured effects were generally smaller than those reported by the original studies.

The probability that a positive research finding is a true-positive result is known as the **positive predictive value (PPV)**. This depends on the

Box 2.2 Lack of replication due to chance when effects are weak and samples small

Assume that a coin is weakly biased and has a probability of 0.6 of landing heads up (compared with 0.5 for a fair coin). The hypothesis that the coin is biased is tested in two experiments, with the following results:

Experiment 1: You toss the coin six times and obtain six heads.
Experiment 2: You toss the same coin six times and obtain three heads and three tails.

If the coin was fair, the probability of getting six heads from six tosses, as in Experiment 1, would be only 0.0156. As this probability is lower than 0.05, which is the conventional criterion for statistical significance, you correctly reject the null hypothesis that the coin is fair and conclude that it is biased – in this case, a true-positive result. However, you then conduct Experiment 2 and get a different result. The second experiment alone provides no evidence that the coin is biased – a false-negative result. It turns out that the outcome of Experiment 1 was quite unlikely given how weakly the coin is biased: the probability of getting six heads with the biased coin is only 0.0467 (~5 per cent). Although the conclusion from the first experiment was correct, the observed result was lucky, given the weak bias of the coin. With experiments consisting of six tosses, only about one in 20 experiments (5%) will correctly reject the null hypothesis that the coin is fair. This is an example of a massively underpowered experiment: six tosses are way too few to reliably reach the correct conclusion that the coin is biased. In fact, an experiment with this coin would need 169 tosses to have an 80 per cent chance of correctly rejecting the null hypothesis. We return to the issues of power and sample size in Chapter 4.

probability of the finding being true *before* the study was conducted, the statistical power of the study and the threshold adopted for claiming statistical significance (typically $p < 0.05$). Higher power results in higher PPV. A fact that is not widely appreciated is that positive results are more likely to be false positives in studies that are underpowered and therefore have low PPV (Box 2.3) [22]. It follows that positive results obtained from studies with small sample sizes should be regarded with greater scepticism.

In summary, low-powered studies with small sample sizes are more likely to produce unreliable findings for three reasons: the probability of false-

---

Box 2.3 Positive results are less likely to be true positives in studies with small sample sizes

Returning to the example started in Box 2.2, assume that in the population of all coins, 20 per cent are biased, with a probability of 0.6 of landing heads up, and 80 per cent are fair. If you pick a coin from this population at random, the probability of the coin being biased *before* testing it in an experiment is 0.2. You now pick a coin at random from the population and conduct two experiments to test whether it is biased.

Experiment 3: You toss the coin six times and obtain six heads.
Experiment 4: You toss the coin 100 times and obtain 59 heads.

What can you conclude from these experiments? Just as in Experiment 1, standard statistics can be used to reject the null hypothesis that the coin you picked is fair. You therefore conclude in both experiments that the coin is biased. But how likely is it that these two positive results are true? In the light of the information you have about the frequency of biased coins in the population, you can calculate the PPV for Experiments 3 and 4 as follows:

$$PPV = (R[1 - \beta])/(R[1 - \beta] + \alpha)$$

where R is the pre-study odds that an effect is non-null, $[1 - \beta]$ is the power of the study and α is the criterion for statistical significance. Note that the odds are related to the probability (p) that the finding is true before doing the study as follows: $R = p/[1 - p]$. Thus, R for Experiments 3 and 4 is 0.25.

For Experiment 3 with a sample size of six, the PPV is only 0.19, whereas for Experiment 4 with a sample size of 100 the PPV is a much higher 0.76. Although both experiments provide evidence to reject the null hypothesis that the coin is fair, you can be much more confident that the result obtained in Experiment 4 reflects the truth.

---

negative results is higher; the probability that positive results reflect true-positive results is lower; and estimates of the size of true-positive effects are exaggerated.

The broad conclusion from many replication studies is that the original reports of some scientific findings are false-positive results and the subsequent failures to replicate these findings are consequently true negatives. The main reason is that the original studies were typically based on small sample sizes.

2.3.3 Dubious Research Practices

While sample size is undoubtedly important, small, underpowered experiments are not the only explanation for the replication crisis. Various other factors have contributed to the prevalence of false-positive results in the published literature. These range from biases in the processes of publication, through dubious research practices to downright fraud.

Publication bias is the tendency for positive results to be published and negative results to remain unpublished. It has undoubtedly made a large contribution to the replication crisis. Negative results often fail to make it into the published literature – a phenomenon sometimes referred to as the 'file drawer' problem. The problem is that researchers and journal editors prefer tests of hypotheses that are novel or that challenge current views. Providing evidence for obvious hypotheses or null hypotheses has rarely made academic careers or sold journals.

One revealing finding to emerge from replication studies is that researchers' beliefs about how likely the hypothesis being tested is to be true are themselves a good predictor of whether a study will confirm the original finding. Quite simply, results considered to be implausible are less likely to be confirmed by successful replication [13]. This makes sense because, as we saw above, the PPV is reduced when the prior probability of a positive result being true is low. Consequently, the kinds of positive findings most favoured by researchers and journals are exactly those that are least likely to be true and therefore the least likely to be replicated.

One solution to this problem is adopting different criteria for statistical significance, depending on the type of study. A suggestion endorsed by many researchers is that while the conventional level of $p < 0.05$ is appropriate for confirmatory or contradictory replications, a more stringent criterion of $p < 0.005$ should be used for the discovery of new effects [23]. This would have the desired consequences of reducing the false-positive error rate and increasing the probability that new discoveries are actually true.

It is worth emphasising that in the context of scientific findings, 'positive' and 'negative' have no absolute meaning. The valence (direction) of a result is defined relative to the hypothesis under test. Showing that peahens do not prefer males with more elaborate trains is only a negative finding if the hypothesis being tested was that they would. Part of the solution to the replication crisis would be to publish findings whatever their direction. All else being equal, a strong negative test of an implausible hypothesis should be accorded the same value to science as a positive test of the same hypothesis.

A second reason why there are so many false-positive results is that researchers may consciously or unconsciously exploit flexibilities in data collection, analysis and reporting (researcher degrees of freedom) in order to produce evidence for a desired positive result. This flexibility increases the potential for turning true-negative results into false-positive results [24]. Study conditions may be reported selectively. Data collection may be stopped once a statistically significant difference emerges. Dependent variables may be redefined or switched, outlying data points removed and data dredged until significant findings emerge – practices known collectively as *p*-hacking. Negative results may be actively suppressed and post-hoc hypotheses constructed to fit the actual findings – another improper practice known as **HARKing** (Hypothesising After Results are Known).

The way in which individual scientists are evaluated on the basis of their 'high-impact' publications creates incentives for ambitious researchers to cheat in these ways, whether deliberately or unintentionally. In anonymous surveys, only 2 per cent of researchers admitted to having actually fabricated results [25]. However, exploiting the flexibilities listed above was a much more common offence, with 34 per cent of researchers admitting to such questionable practices [25]. The true percentage is probably higher because it is often possible to justify these practices as legitimate, both to oneself and to journal reviewers and editors. In the following section, we explore some procedures that have been introduced to drive out dubious research practices and make behavioural science more replicable.

In summary, threats to replication arising from low power and dubious practices afflict all stages of the research cycle, from hypothesis generation through to publication (Figure 2.1).

2.4 Identifying Reliable Sources of Evidence

One of the most important take-home messages from the replication crisis is that the stated findings of a single, published study should not simply be accepted at face value. Behavioural scientists should assess the quality of the evidence for a given finding. How can they do this?

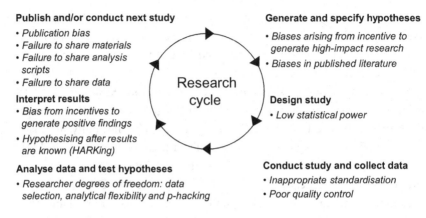

Publish and/or conduct next study
• Publication bias
• Failure to share materials
• Failure to share analysis scripts
• Failure to share data

Interpret results
• Bias from incentives to generate positive findings
• Hypothesising after results are known (HARKing)

Analyse data and test hypotheses
• Researcher degrees of freedom: data selection, analytical flexibility and p-hacking

Generate and specify hypotheses
• Biases arising from incentive to generate high-impact research
• Biases in published literature

Design study
• Low statistical power

Conduct study and collect data
• Inappropriate standardisation
• Poor quality control

Research cycle

Figure 2.1 Threats to replication of science. The various stages of the scientific method are shown in bold and the various threats to replication present at each stage are listed. Modified from [26].

2.4.1 The Evidence Pyramid

An important idea that was developed in the context of evidence-based medicine is that of a hierarchy of scientific evidence, whereby some types of studies are weighted more highly than others in establishing the truth. The hierarchy is often illustrated with a pyramid, which places the strongest types of evidence at the top and the weakest at the base. A modified version of the evidence pyramid, which is broadly applicable to the behavioural sciences, is shown in Figure 2.2. The pyramid can be used to evaluate the quality of the evidence that a predictor variable X is responsible for causing changes in an outcome variable Y.

The base of the pyramid (layer 5) consists of anecdotes, ad hoc personal observations and personal communications from other researchers. Such observations have little evidential value because they are typically based on small sample sizes and are not recorded or reported systematically. Even so, they can contribute to the development of new hypotheses and stimulate further research.

The next layer up (layer 4) comprises qualitative studies. These have the advantage of following established methodologies and collecting data on multiple subjects or instances of behaviour. Qualitative studies can provide evidence that an association might exist between two variables and give

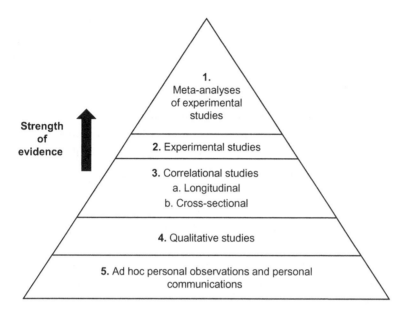

Figure 2.2 An evidence pyramid for the behavioural sciences.

clues as to possible causal relationships. Data from qualitative studies has value for hypothesis generation and as a stimulus for further research. Its main limitation, however, is that it does not allow statistical comparisons between groups or quantitative synthesis of evidence across studies.

Sitting above qualitative studies in the pyramid are **cross-sectional** correlational studies (layer 3b). These are observational studies that measure the association between two variables at a single point in time. Their main shortcoming is that they cannot prove causation. A correlational study can show that X and Y are significantly correlated, but from correlational data alone it is impossible to be sure whether X causes Y, Y causes X, or some third variable causes both X and Y. **Longitudinal** correlational studies (layer 3a), where subjects are measured more than once over time, can provide stronger evidence for causation. If a change in X over time is correlated with a change in Y, it becomes more likely that there is a causal relationship between X and Y. However, just as with cross-sectional correlational studies, it is still not possible to rule out third-variable explanations.

Experimental studies (layer 2) provide the strongest form of evidence available from a single study that X causes Y. In a well-designed experiment, randomly chosen subjects are randomly allocated to different levels (or conditions) of the independent variable X (typically experimental and control conditions) and the dependent variable Y is then measured. Experimental studies are the only type of study that can conclusively test hypotheses about causation because third-variable explanations are eliminated by random assignment of subjects to conditions.

At the top of the evidence pyramid (layer 1) are **meta-analyses,** which employ statistical methods for synthesising results from multiple studies [27]. The component studies are generally identified by a systematic review procedure that makes their selection fully transparent. Meta-analyses of correlational studies provide the strongest evidence about associations (X is correlated with Y) and meta-analyses of experimental studies provide the strongest evidence for causation (X causes Y). Meta-analyses provide more powerful estimates of true effect size than those derived from single studies and help to address the problem that individual studies may report false-positive or false-negative findings. An important caveat is that the evidence on which meta-analyses are based will be biased if, as is known to be the case, negative results are less likely to be published than positive results.

Within each layer of the pyramid, studies of the same type provide stronger evidence when they are based on a larger sample size. Within layer 2, for example, a full-scale experiment with 100 subjects provides stronger evidence than a pilot experiment with five subjects. As mentioned earlier, one of the most important lessons to emerge from the replication crisis is that sample size is the single most important predictor of whether a finding will be replicated.

2.4.2 Different Types of Publication

Where a study is published can indicate the likely quality of the underlying evidence and the confidence that can be placed in its findings. Studies published in **peer-reviewed** scientific journals have been reviewed by at least one, and usually two or three, academics in the field before publication. Journals also typically publish information about their editorial standards and peer-review procedure. In contrast, preprints or other types of non-peer-reviewed publications may not have undergone such careful scrutiny and are more likely to contain methodological errors and unsubstantiated claims. The **grey literature** is a term used to refer to research produced by organisations or individuals outside the traditional academic publishing channels, and includes working papers, reports and government documents. Some grey literature is of high quality, but the standard of evidence should be scrutinised more carefully.

Scientific journals are rated according to their **impact factor**, a statistic based on how frequently the papers published in that journal are cited in other publications. The journals with the highest impact factors are widely regarded as the most prestigious and *should* contain the best-quality science, reflecting the strong competition to publish in them and their rigorous selection and review processes. That said, findings published in the highest impact journals are also the most likely to be retracted [28, 29]. **Retraction** occurs when the findings of a paper are no longer considered trustworthy because of errors or scientific misconduct of various kinds. The higher retraction rate among high-impact journals probably arises because their editors have a bias towards publishing positive tests of implausible hypotheses, which are the most likely to yield false-positive findings that later turn out to be untrue [30]. Many reputable journals now subscribe to a system (called CrossMark) whereby a button on the electronic version of each paper gives easy access to its current status, including whether the paper has been updated, corrected or retracted since publication.

2.4.3 The Open Science Movement

The prevalence of dubious research practices highlighted by the replication crisis has given rise to a number of practical innovations designed to strengthen the quality of science and the confidence in published findings. 'Open Science' is an umbrella term to describe these practices [26–31].

The problems arising from researcher degrees of freedom are addressed through various forms of study **preregistration**. Preregistration involves logging a statement of a study's research question, hypotheses and predictions, along with a description of the data to be collected and the statistical analyses that will be used to test the hypotheses, *before* data collection is started. A common criticism is that preregistration limits creativity by preventing researchers from exploring their data in ways they had not previously anticipated. This is a misconception: preregistration does not

---

Box 2.4 Questions that need to be addressed in a basic preregistration (adapted from AsPredicted.org)

1. **Data collection**. Has any data been collected for this study already?
2. **Hypothesis**. What is the main question being asked and hypothesis being tested?
3. **Outcome variable**. What are the key outcome (dependent) variable(s) and how will they be measured?
4. **Predictor variable(s)**. What are the key predictor variable(s)? If relevant, what treatments and levels will subjects be assigned to?
5. **Analyses**. What analysis will be conducted to test the main hypothesis?
6. **Outliers and exclusions**. If relevant, how will outliers be defined and what rule(s) will be used for excluding data points?
7. **Sample size**. How much data will be collected and how has the sample size been determined?
8. **Other analyses**. Are any secondary analyses or exploratory analyses planned?

---

preclude exploratory analyses [32]. Rather, it forces researchers to be transparent about which analyses were pre-planned and which were conceived after the event. Although exploratory analyses can be mentioned in a preregistration, there is no requirement that exploratory analyses must be preregistered. Various web platforms have been developed to host preregistrations, such as the Open Science Framework and AsPredicted.org, and preregistration is rapidly spreading in the behavioural sciences. A list of questions that need to be addressed in a basic preregistration is given in Box 2.4.

A growing number of science journals have taken preregistration a step further by offering what are called **registered reports**. Under this format, the researcher writes the introduction and methods for their proposed study and submits them to the journal for peer review. If the journal accepts the registered report, it also commits to publishing the final paper regardless of whether the results are positive or negative. Registered reports therefore tackle problems arising from researcher degrees of freedom and publication bias. The Open Science movement has also fostered a shift towards encouraging replication studies. Many academic journals now explicitly state that they accept replications.

Even though transparency over methods and open sharing of data are core values in science, they have not always been common practice. The Open Science movement has sought to change this by actively encouraging the sharing of materials, analysis scripts and data. Many journals now require that materials and data are made publicly available at the time a paper is published. Complete transparency over data manipulation and statistical analysis can be achieved by making available the raw data from a study alongside an analysis script that generates all the statistics in the publication. Such scripts can be written in a freely available statistical programming language such as R. Just as with preregistration, various web platforms have emerged that host datasets and associated analysis scripts (e.g. Open Science Framework, Dryad and Zenodo). Many of these platforms provide a **doi** (digital object identifier) – a standardised, persistent handle used to identify digital objects – allowing proper citation of data files by researchers who want to reuse them.

The TOP (Transparency and Openness Promotion) guidelines [33] are intended to create shared standards for Open Science practices across academic journals. These guidelines include standards relating to replication, preregistration and transparency over materials, methods and data. Journals can decide which tier of standards to adopt. For example, in relation to data transparency, the lowest tier requires that a paper includes a statement about data availability; the next tier up requires that the data is made publicly available as a condition of publication; and the highest tier requires third-party verification before publication that the data is accessible and useable.

2.5 Summary

- Behavioural studies aim to discover scientific truths.
- True facts should be replicable, meaning that the same conclusions are reached if the same data is analysed, if the same methods are applied to collect a new dataset and if different methodological approaches are used to address the same general hypothesis.

- The replication crisis refers to a widespread failure to replicate published findings in the biological and social sciences.
- The causes of the replication crisis include the presence of uncontrolled moderators of behaviour, low statistical power and dubious research practices.
- Various sources of information can help to distinguish good research from bad. An evidence pyramid ranks different study types according to the quality of evidence produced.
- The Open Science movement encourages replication, preregistration and transparency over materials, methods and data, all of which should improve the quality of science and the likelihood that findings will be replicated.

3
Choosing a Research Question

Charles Darwin exclaimed: 'How odd it is that anyone should not see that all observation must be for or against some view if it is to be of any service!' Darwin was right that all scientific research should start with a question. A study is unlikely to be fruitful if it never focuses on specific questions.

Setting out to measure behaviour without a specific question in mind raises at least two problems. First, behaviour is complex and it is impossible to measure all aspects with the same accuracy and precision. To collect good data, researchers must know where to focus their effort. Second, even when it is possible to measure multiple aspects of behaviour well, the more things that are measured, the higher the risk of getting false-positive results by chance. It follows that the results obtained from unfocused research, in which many different variables are measured without knowing why, are less likely to be true.

A clear research question directs and focuses the whole research cycle. Once identified, the question leads naturally to hypotheses; and hypotheses in turn lead to predictions about what the data should show, given specific observations or experiments. The predictions shape the choice of methods for data collection and statistical analyses. Finally, the exercise of posing and answering a specific question provides the framework for reporting the study and should shape the structure of the resulting scientific paper. It is therefore critical to keep the research question clearly in mind at all stages. If you find yourself getting bogged down or side tracked in the course of a study, it is useful to remind yourself what question you originally set out to answer. The importance of being clear about the research question further reinforces the value of preregistering a study.

Given the central role of the research question in conducting good scientific research, it is obviously important to choose it carefully. In this chapter, we explore some of the issues that should be considered when choosing a research question and generating hypotheses and predictions from it.

3.1 Tinbergen's Four Questions

A number of fundamentally different types of question can be asked about any behaviour. A useful and widely accepted classification was formulated in the 1960s by the Nobel Laureate ethologist Niko Tinbergen, who pointed out that four logically distinct types of question are raised by the study of any behavioural trait [34, 35]. They are:

Mechanism, or how does it work? This is also sometimes referred to as the question of proximate causation. Questions of mechanism concern the internal and external causal factors that elicit and control the behaviour from moment to moment. These causal factors include physical and social stimuli from the environment and the underlying psychological, physiological and neurobiological mechanisms that regulate an organism's behaviour.

Development, or how did it develop? This is also sometimes referred to as the question of ontogeny. Questions of development are about how the behaviour is assembled and how it changes during the lifetime of the individual, from conception to death. As with mechanism, this involves understanding how external and internal factors influence the way in which the behaviour develops. In the case of development, however, the focus is on the interplay between the current state of the individual and its genes and past environment. Questions of behavioural development include the role that different genotypes and different types of experience may play in altering the developmental process.

Function, or what is it for? This is also sometimes referred to as the question of current utility or adaptive significance. Questions of function concern the current use of the behaviour in helping the individual to survive and reproduce in its current physical and social environment. When the focus is on function, there is no presumption about the historical processes that gave rise to the behaviour, either during the lifetime of the individual or over the evolutionary history of its species. The current function of a behaviour need not equate with its original function because behaviours can change their function over time. Current function need not be the product of natural selection and may arise through individual learning and cultural processes.

Evolution, or how did it evolve? This is also sometimes referred to as the question of phylogeny. Questions of evolution concern how the behaviour arose during the evolutionary history of the species. Evolutionary questions encompass the evolutionary process and the factors involved in moulding the behaviour over the course of evolutionary history. They are distinct from questions of current function, although the two are sometimes muddled.

Tinbergen's four questions can be illustrated with a simple example. Suppose we ask why it is that drivers stop their vehicles at red traffic lights. An answer in terms of mechanism could be that a visual stimulus (the red light) is perceived via the eyes, processed in the central nervous system and reliably elicits a specific motor response (easing off the accelerator, applying the brake and so on). An answer in terms of development could be that individual drivers have learned this rule from the media and driving instructors. An answer in terms of function is that drivers who do not stop at red traffic lights are liable to crash or be stopped by the police. Finally, an 'evolutionary' answer would deal with the historical processes whereby a red light came to be a universal signal for stopping traffic at road junctions across modern human cultures. All four answers are equally valid, but they represent four distinct types of enquiry about the same behavioural phenomenon. Although Tinbergen was mainly concerned with the behaviour of humans and non-human animals, his four questions are equally applicable to the behaviour of microbes, plants and even AI agents [2].

Tinbergen's four questions can be organised into a two-by-two table according to the object of the explanation and the kind of explanation sought (Table 3.1) [36]. The object of explanation can be either the historical sequence that results in the behaviour (development and evolution) or

Table 3.1 Tinbergen's four questions

| | | Object of explanation | |
		Historical sequence	Single time point
Kind of explanation	Proximate (how?)	Development (ontogeny): how did it develop?	Mechanism: how does it work?
	Ultimate (why?)	Evolution (phylogeny): how did it evolve?	Function: what is it for?

the behaviour at a single point in time (mechanism and function). The kind of explanation sought can be either proximate (development and mechanism) or ultimate (evolution and function).

A failure to distinguish between Tinbergen's four questions lies at the heart of many heated debates in the behavioural sciences. For instance, consider the relatively high rates of teenage pregnancy in the UK. When asked to explain this phenomenon, a sociologist might argue that teenage pregnancy is caused by exposure to an environment of economic and social deprivation. A physiologist, on the other hand, might argue that it is caused by accelerated maturation and the early onset of puberty, while an evolutionary psychologist might argue that teenage pregnancy is an adaptive biological response to reduced healthy life expectancy. But these are not mutually exclusive explanations. Rather, they are different *types* of explanation, each addressing a different one of Tinbergen's four questions. The sociologist is suggesting a mechanistic answer, the physiologist a developmental answer and the evolutionary psychologist a functional answer. All three explanations could be correct. Unfortunately, explanations in terms of mechanism, development, function and evolution are often pitted against one another, as though they were mutually exclusive, whereas they are equally valid answers to different questions.

Although Tinbergen's four questions are logically distinct, it is sometimes easier to answer one question if you know something about the answers to one or more of the other questions. For instance, many experiments use the tactic of testing a behaviour in an unnatural situation for which it is not adapted (something that can only be done if the function of the behaviour is understood). Behaviour in such unnatural situations can sometimes reveal the mechanisms involved. For example, researchers have unravelled the cues used by migrating birds to determine their heading by examining their behaviour in unnatural situations, in which putative navigational cues (e.g. the direction of the sun, the earth's magnetic field and familiar landmarks) are artificially manipulated to provide conflicting sources of information.

3.2 Different Levels of Causal Explanation

Most behavioural research addresses questions of mechanism (proximate causation). However, the level of organisation at which mechanisms are studied varies depending on the discipline and its concerns. A sociologist might be interested in the social causes of a human behaviour; a psychologist might be interested in the role of psychological constructs such as motivations or emotions; a behavioural physiologist might be interested in hormones mediating the behaviour; and a neurobiologist might be interested in identifying the neural networks underlying it. All of these scientists are interested in mechanism, but they approach the question at different levels of organisation.

The existence of such different approaches to the study of mechanism raises the question of which, if any, is best? Are analyses of the societal determinants of a behaviour more or less valuable than analyses of its underlying physiology or neurobiology? The answer depends on the reason for undertaking the research. If the aim is to inform the development of evidence-based policy on the education of children, for example, then understanding social causes would probably be most useful. If, however, the aim is to develop a new drug to treat a behavioural problem, such as attention deficit hyperactivity disorder (ADHD), then understanding the underlying biological mechanisms would be more relevant. This is why different disciplines study the mechanisms of behaviour at different levels of organisation. The differing approaches provide complementary information about the mechanisms responsible for the behaviour in question.

Understanding the causal mechanisms underlying behaviour at the lowest level of organisation, such as neurons or computer algorithms, does not make it easy, or even possible, to predict how the whole organism or AI entity will behave at the macroscopic level. The converse is also true: in the field of AI, as in behavioural biology, it is generally difficult or impossible to reverse-engineer the entity and explain in purely mechanistic terms why it has behaved in particular ways in different situations [37]. The result is that scientists cannot always explain the behaviour of some AI agents, despite the fact that they themselves designed the underlying computational mechanisms – an issue sometimes referred to as the problem of explainability. Some scientists argue that the behaviour of AI systems should be studied as though they were biological organisms whose neural mechanisms are unknown [2].

3.3 Choosing the Right Subjects

Identifying the basic research question is just the start. The next step is to consider which population of subjects to study. For many scientists, choosing which species to study is not an issue. They may only be interested in humans; they may want to study one species because it is endangered; they may want to understand an abnormal behaviour that is most easily studied in a particular species; or they may just have to study whatever species is available in their laboratory. Nonetheless, when choice is possible, it is worth considering the attractions and pitfalls presented by the vast range of species that could be studied.

Choosing which species to study is likely to be more of an issue for researchers interested in a type of behaviour that is present in many different species, such as how male mating displays affect female choice, why animals perform stereotyped movements when confined to small enclosures, or how auditory experience affects the development of birdsong.

When several candidate species display the behaviour in question, the challenge is choosing the best one to study. Relevant questions to consider might include the following. Does an extensive biological and behavioural literature exist for this species? Is its natural behaviour well suited to the particular problem to be investigated? How solitary or gregarious is it? What are its life-history characteristics, such as gestation period, age of independence and age of sexual maturity? Is its life-span long enough to make repeated measurements possible but short enough to make studies of development practicable? Has its genome been sequenced? If the animals have to be imported, are the conditions for collecting them in the country of origin ethical? Does the species respond well if kept in captivity or hand-reared? Does it breed successfully in captivity? Does it have problematic dietary or other requirements?

Another situation in which the choice of species becomes an issue is when the researcher is interested in human behaviour and wishes to find a good **animal model** in which to study it experimentally – for example, to investigate such things as the influence of parental neglect on the development of adult anxiety or the effects of an experimental drug on short-term memory. Different species may differ in how closely they resemble the human condition being modelled. Choosing a good animal model requires a thorough understanding of the similarities and differences between species.

Even when the choice of species has been determined, there may still be decisions to make over which population to study. Different populations within a species sometimes differ markedly in their behaviour. For example, the behaviour of standard laboratory mice is different from that of wild members of the same species. The results obtained from one population will not necessarily generalise to other populations of the same species, and this limitation should be highlighted when the results are communicated. If the results are intended to be relevant to a specified population, then it is necessary to study a representative sample of that population. When studying humans, well-developed methods are available for obtaining a representative sample that mirrors the target population in terms of age, gender, ethnicity, socio-economic status and so on. A valid criticism of much human behavioural research is that it is disproportionately based on studies of so-called **WEIRD** subjects (Western, Educated, Industrialised, Rich and Democratic) [38].

3.4 Generating Research Questions

The first step in identifying a good research question is to search and read the relevant scientific literature. The would-be researcher first needs to understand whether other researchers have already asked the same question and, if so, what answers they have found.

Research questions tend to become more specific as more is discovered about a particular phenomenon. It is normal to start with a rather general question and focus in as more is discovered. If little is known about a phenomenon, then it may be better to start with very simple questions, such as the circumstances in which the behaviour occurs, before digging down into more detailed questions about its neurobiological mechanisms.

At the start of a new piece of behavioural research, systematic measurement should be preceded by a period of informal observation aimed at understanding both the subjects and the behaviour to be measured. Preliminary observation is important for two reasons: first, because it provides raw material for formulating better research questions and hypotheses, and second, because choosing the right metrics and recording methods requires familiarity with the subjects and their behaviour. Preliminary observation is especially important if the problems or subjects are new to the researcher.

Beginners sometimes falter early in a study because they rush to collect 'hard data' and do not allow sufficient time to watch, think and frame better questions. Even experienced observers need to spend time on preliminary observation. After a period of trial recording sessions, in which behavioural categories and measurement techniques are refined, preliminary data should be analysed. It is at this stage that hypotheses, predictions and methods should be modified if necessary.

The successful scientist is likely to be one who combines a purposeful approach to tackling a question with the ability to recognise and respond opportunistically to new questions and opportunities that arise during the course of research. If one problem is pursued in a rigid and inflexible way to the exclusion of all else, then observations leading to new ideas may be missed.

The former US Secretary of Defense Donald Rumsfeld once stated: 'There are known knowns. There are things we know that we know. There are known unknowns. That is to say, there are things that we know we don't know. But there are also unknown unknowns. There are things we don't know we don't know.' Most conventional scientific research is about investigating known unknowns. Researchers formulate and test hypotheses according to the standard scientific approach shown in Figure 2.1, which is referred to as the **hypothetico-deductive model**. The hypotheses embody a set of possibilities about what might be found that have emerged from existing theory and data. This approach is referred to as **confirmatory analysis** because it involves the confirmation (or rejection) of *stated* hypotheses. Occasionally, however, a result emerges that is unexpected and not predicted by any of the hypotheses being tested – it is an unknown unknown. The search for unknown unknowns in data is referred to as **exploratory analysis**.

The value of exploratory analysis is that it can yield unexpected results that may change the direction of research and even bring about **paradigm shifts** in scientific understanding, whereby a step change occurs in understanding. The danger of exploratory analysis is that it is more likely to yield false-positive results. For this reason, it is extremely important to distinguish clearly between confirmatory and exploratory analyses. Unexpected results discovered through exploratory analysis must be verified in subsequent confirmatory research.

3.5 Hypotheses and Predictions

Once the basic research question is chosen, the next step is to formulate a set of hypotheses. Formulating hypotheses is – or should be – a creative process, requiring imagination as well as knowledge of the issues involved. In some cases, a considerable amount of descriptive information may be required before useful and interesting hypotheses can be formulated, whereas other problems lend themselves more easily to setting out specific hypotheses at an early stage.

Where previous researchers have suggested explanations for the phenomenon of interest, it may be helpful to think about potential weaknesses in their explanations as a stimulus to formulating new hypotheses. It may also be worth considering hypotheses that have been successful in adjacent areas of research, to see if they can be adapted to the area under investigation. For example, in the social science literature, the observed positive association in humans between obesity and food insecurity (the inability to predictably access sufficient food) is often described as paradoxical. How can it be that people who are sometimes hungry are fatter than those who have secure food supplies? However, the observation makes perfect sense to behavioural ecologists studying weight regulation in small birds because animals lay down extra fat as insurance against starvation if their food supplies are unpredictable [39]. Thus, an observation that seems mysterious in one discipline is well understood in another. Seeking out related research in other fields can accelerate the progress of science.

Within reason, the larger the number of competing hypotheses, the better. The danger with having only one hypothesis is that the researcher may find it harder to abandon their cherished hypothesis if its predictions are not supported by the evidence. Moreover, finding empirical support for a single hypothesis does not prove that it is the only possible explanation. When considering a range of potentially competing hypotheses, it is important to consider whether they are mutually exclusive alternatives, or whether more than one hypothesis could be true (Box 3.1). The aim should be to find the best explanation for the observed data. That means not just any old explanation that is compatible with the data obtained so far, but one that unifies and offers a common account of superficially diverse phenomena.

All hypotheses have a corresponding **null hypothesis**, which is simply the hypothesis that whatever is stated in the main hypothesis (also known as the

Box 3.1 Example of a research question, hypotheses and associated predictions

Some bird species, such as the European starling, collect and place leaves or sprigs of fresh green plants in their nests. This green material (GM) is not part of the basic nest structure, raising the question of its possible function. In European starlings, only males place GM in nests and they do this prior to the start of egg-laying.

Research question: What is the function of the behaviour whereby male starlings place GM in their nests during the period prior to the start of egg-laying?

Hypothesis 1: courtship. Male starlings place GM in nests to attract females.

Null hypothesis 1: GM does not attract females to nests.

Prediction 1.1: Unpaired males are more likely to place GM in nests than paired males.

Prediction 1.2: Males are more likely to place GM in nests in the presence of females.

Hypothesis 2: nest protection. GM reduces parasite or pathogen levels in the nest due to chemical compounds present in the specific GM chosen.

Null hypothesis 2. Chosen GM does not reduce parasite or pathogen levels.

Prediction 2.1: The GM chosen by starlings is a non-random selection of available plants that have antiparasitic or antimicrobial properties on parasites or pathogens known to affect starlings.

Prediction 2.2: Nests with experimentally reduced GM have increased parasite or pathogen levels.

Hypothesis 3: drug effects on nestlings. Compounds present in GM directly improve the health and development of nestlings.

Null hypothesis 3. GM has no direct effect on nestling health and development.

Prediction 3.1: Experimentally increasing GM in nests improves nestling health and growth without reducing nest parasite or pathogen levels.

Note that, in this case, the three hypotheses are not mutually exclusive and could all be true. As it turns out, there is empirical evidence to support all three hypotheses [40], illustrating the importance of considering and testing multiple hypotheses.

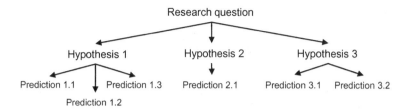

Figure 3.1 The hierarchy of a research question with multiple hypotheses, each of which makes one or more testable predictions. Each hypothesis will have a corresponding null hypothesis.

alternative hypothesis) is not true. Science and statistical tests work by testing null hypotheses. If the evidence allows the null hypothesis to be rejected, then it supports the alternative hypothesis. Thus, science proceeds through the conservative approach of assuming that nothing interesting is happening (i.e. that the null hypothesis is true) unless there is evidence to the contrary.

Once the hypotheses have been formulated, the next step is to work out what predictions flow from them. Predictions must be specific and testable. The more specific the predictions, the easier it usually is to distinguish between competing hypotheses and thereby reduce the number of different ways in which the results could be explained. At least one prediction is needed for each hypothesis, but making several different predictions from a hypothesis will produce a stronger test (an approach known as constructive replication or triangulation – see Chapter 2). At this preliminary stage, it is worth sketching graphs of how the data might look if the prediction is supported and, conversely, how the data would look if the prediction fails and the null hypothesis turns out to be true. Doing this helps to clarify how the data will be analysed and used to test the predictions (see Chapter 11).

The relationship between a research question, a set of hypotheses and the testable predictions arising from each hypothesis can be depicted as a hierarchy of increasing specificity (Figure 3.1). Box 3.1 provides a real example.

3.6 Summary

- Identifying a good research question is a vital first step in any behavioural study because the question will focus the rest of the research cycle.

- Four logically distinct types of question can be asked about any behaviour. These concern its mechanisms, its development (or ontogeny), its function and its evolution (or phylogeny).
- The mechanisms underlying behaviour can be studied at many different levels, ranging from the social or physical environmental conditions that influence the behaviour down to the neural networks responsible for behavioural output.
- The nature of the research question will influence decisions about what species to study.
- Research questions are developed through a combination of approaches, including reading the literature, preliminary observations and exploratory data analysis.
- A research question leads to a set of hypotheses that need not be mutually exclusive but should all be testable. Each hypothesis should generate one or more specific predictions.

4
Designing a Behavioural Study

The aim of much behavioural research is to distinguish between competing hypotheses for how behaviour can be explained. As we saw in Chapter 3, this involves testing specific predictions, but coming up with testable predictions is only the first step. Planning good behavioural research involves making a number of decisions about exactly how the predictions will be tested. This means designing a study with the right structures and processes.

A poorly designed study will waste time and resources and, at worst, produce results that are misleading or uninterpretable. While it is sometimes possible to rescue useable data from poorly designed studies by means of clever statistical analysis, it is far better to arrive at the right design before collecting the data. Study design, which includes experimental design, is a big topic, and many specialist text books are available [41–43]. Our aim here is to give an overview of the main issues, highlighting those that apply especially to behavioural research.

4.1 Aims of Good Study Design

Living organisms are variable, and their behaviour is particularly so. When researchers test a prediction about behaviour, they are seeking to explain some of this variation. But even if they are successful, they are unlikely to explain all of the observed variation. There will generally be some additional, unexplained **random variation** resulting from the effects of uncontrolled or unmeasured variables, individual differences and measurement error. One major aim of good study design is to minimise random variation. Good study design is about maximising the signal-to-noise ratio, where the signal is the effect the researchers are trying to detect and the noise is the random variation that obscures the picture.

Another fundamental aspect of behaviour is that its causes are multifactorial. An organism's behaviour is the result of complex interactions between its genes, its past experience and its current environment. If the aim is to test the effect of variable X on behaviour Y, and variable Z also affects X and Y, then Z is a potential **confounding variable**. For example, imagine that researchers wanted to test the prediction that laboratory mice handled by the tail become more anxious than mice handled by being coaxed into a small plastic tunnel. An experiment is performed in which mice are randomly assigned to be either tail-handled or tunnel-handled daily for 2 weeks. Their anxiety levels are then assessed using a standardised open-field test, which involves measuring how much time the mouse spends near the walls when it is placed in an open arena. Suppose, however, that the researchers assess all the tail-handled mice in the morning and all the tunnel-handled mice in the afternoon. In this study design, the handling manipulation is perfectly correlated with the time of the measurement and the two factors are said to be confounded. Consequently, it is impossible to tell whether any difference in anxiety is due to the handling method or the time of testing. This would clearly be a poorly designed experiment. Many behaviour patterns change according to time of day, making time a potential confounding variable in behavioural studies. The time at which measurements are made therefore needs to be carefully planned (see section 4.8). A well-designed study should eliminate or control for potential confounding variables such as this. In the following sections, we describe different types of study design and outline strategies to reduce the effects of random variation and eliminate confounding variables.

4.2 Different Types of Design

Perhaps the most important distinction in study design is between **correlational studies** and **experiments**. Correlational studies – also known as observational studies – make use of naturally occurring variation to test predictions about the likely causes of variation in behaviour, whereas in experiments the researcher manipulates some aspect of the situation to create artificial variation and then measures the effects of this manipulation on behaviour. In many cases, it will be possible to test predictions about behaviour using either approach, but they have different advantages and disadvantages.

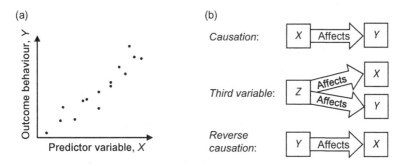

Figure 4.1 (a) Scatterplot showing a positive correlation between a predictor variable, X, and an outcome behaviour, Y. (b) Three possible causal explanations for the relationship shown in (a).

At this point, we need to establish some terminology. The term **outcome** refers to the behavioural variable that researchers are attempting to predict. Synonymous terms include **response variable** and **dependent variable**. The outcome variable is usually the one plotted on the vertical (y) axis of graphs (Figure 4.1a). The term **predictor** refers to the variable (or variables) that are predicted to affect the outcome variable. Synonymous terms include **explanatory variable** and **independent variable.** The terms dependent and independent variable are only used in the context of experimental studies. The predictor variable is usually plotted on the horizontal (x) axis of graphs.

4.3 Correlational Studies

The defining feature of a correlational study is that it does not involve any manipulation of the predictor variables. Correlational studies are often appropriate for measuring behaviour in the natural environment, where it may be difficult or unethical to perform experimental manipulations. However, correlational studies are not confined to natural environments; many studies aimed at understanding naturally occurring individual differences in behaviour bring human or animal subjects into the laboratory to assess them under standardised conditions. Suppose, for example, that researchers were interested in testing the prediction that people who have grown up in economically deprived neighbourhoods are more impulsive in their decision making. A study to address this question could involve recruiting participants, giving them a questionnaire to establish where they grew up and then bringing them into the laboratory to assess their impulsivity. Impulsivity could be assessed through a computer game in which the participant makes choices between smaller–sooner and larger–later rewards. In this hypothetical study, the index of deprivation of the childhood neighbourhood is the predictor variable and the proportion of smaller–sooner choices is the outcome variable. Such studies are sometimes mistakenly referred to as experiments, perhaps because they occur in a laboratory, but in fact they are correlational studies because there is no manipulation of the predictor variable.

Correlational studies are usually easier to perform and ethically less problematic than experiments. For these reasons, they are often a good starting point in behavioural research. In some areas of science, such as astronomy and geology where experiments are seldom feasible, correlational studies may be the only option. Detailed quantitative predictions are regularly tested by observation alone.

A major scientific advantage of correlational studies is that because the variation in predictors is within the natural range, the results are likely to provide some insight into the factors that affect behaviour in the natural environment (a quality called **ecological validity**).

A major limitation of correlational studies is their inability to prove that the predictor is responsible for causing the observed variation in the outcome. An observed positive correlation between a predictor variable X and a behavioural outcome Y could have one of three possible explanations (Figure 4.1b): variable X could affect behaviour Y (**causation**); behaviour Y could affect variable X (**reverse causation**); or X and Y could both be affected by a third variable, Z, that has not been considered (**third-variable causation**). Take, for example, the observed correlation between smoking behaviour and accelerated ageing. While the accepted explanation is that smoking causes accelerated biological ageing, it is at least theoretically possible that people who are ageing faster are also more likely to smoke (reverse causation), and there is some evidence to suggest that exposure to adversity early in life both causes accelerated ageing and makes people more likely to smoke (third-variable causation) [44].

Correlational studies can either be **cross-sectional**, whereby each subject is measured only once, or **longitudinal**, whereby each subject is tracked over a period of time and measured on two or more occasions (typically a baseline measurement followed by one or more follow-up measurements). Cross-sectional studies are quicker to conduct but suffer from certain limitations. For example, researchers who were interested in the effects of age on behaviour could perform a cross-sectional study in which subjects of different ages are assessed to see whether they behave differently. If the results show that behaviour is indeed related to age, there could be more than one explanation for this relationship. In particular, it could reflect the fact that older subjects were born at an earlier time when the environment was different. Chronological age might therefore be confounded with some other difference in experience, making it impossible to establish whether age alone was responsible for the observed differences in behaviour. An effect of year of birth is known as a **cohort effect**. Longitudinal studies eliminate cohort effects by looking for effects within subjects rather than across subjects. However, longitudinal studies are harder to perform because individual subjects must be tracked over time and some are likely to be lost to follow-up.

Another problem with cross-sectional measurement is that it combines results from individuals who may be changing in different ways. In Figure 4.2, a behavioural measure (Y) increases sharply over a narrow age range, but the

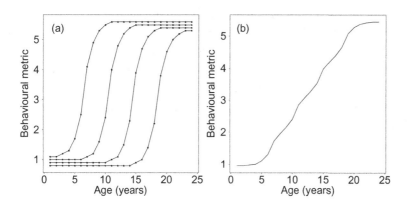

Figure 4.2 A comparison between longitudinal and cross-sectional measurement. (a) The graph shows how a behavioural metric changes as a function of age for four subjects measured repeatedly over time (longitudinal measurements). Each subject exhibits an abrupt increase in behaviour, but the age at which this starts differs between subjects. (b) The graph shows the mean score for the same four subjects. This increases gradually as a function of age, a pattern of development that does not represent any individual subject. A similar picture would have emerged if a different sample of subjects had been measured at each age (cross-sectional measurement).

timing of this increase varies between individuals. A cross-sectional study that measured different subjects at each age would obtain a misleading average score that appeared to show a gradual increase in Y, a pattern of change that would not represent any individual. Gradual learning curves are often an artefact of averaging data from individuals that each show an abrupt transition in behaviour at different time points following the start of training.

Another potential pitfall with longitudinal studies is the cumulative effect of repeated testing. If measurement involves some kind of intrusive behavioural testing, then age-related changes and the experiential consequences of testing will inevitably be confounded. Repeatedly testing a young animal's responsiveness to a stimulus, for example, can influence the course of its development and affect its behaviour in subsequent tests. Even the individual's increasing familiarity with the test situation can affect its behaviour in ways that are not related to any underlying changes in responsiveness.

In summary, cross-sectional and longitudinal approaches raise different problems of practice and interpretation. Ideally, both methods would be used, as exemplified by a classic study of the development of pecking behaviour in domestic chicks. Cruze [45] wanted to measure the effects of age and experience on the accuracy of visually directed pecking. He kept the chicks in the dark from when they hatched and fed them by hand. Starting at different ages, he tested the accuracy with which the chicks pecked at seeds. Once a chick had been tested, it was retested on subsequent days of its life. In this way, Cruze was able to obtain cross-sectional data on chicks that were first tested at a given age, and longitudinal data on chicks that were retested each day. He found that both a chick's age and its prior experience of pecking affected its accuracy.

4.4 Experimental Studies

The defining feature of an experiment is that the researcher manipulates one or more predictor variables while attempting to hold everything else constant. In an experimental study, the manipulated predictor variable is referred to as the **independent variable (IV)** and the measured outcome variable is referred to as the **dependent variable (DV)**. The right design makes it possible to distinguish between the effects of variables that would otherwise be confounded with each other.

The major advantage of experiments over correlational studies is that they allow conclusive inferences to be made about causation. A well-designed experiment eliminates the possibility that an observed correlation between two variables results from reverse causation or third-variable causation.

Experimental studies may be the only option when there is little or no natural variation in the predictor variable of interest. In such cases, it may be possible to create variation artificially by means of an experimental manipulation. An example of this approach is Tinbergen's classic study of egg shell removal in black-headed gulls. These birds invariably remove egg shells from their nest shortly after their chicks have hatched. To understand the biological function of this behaviour, Tinbergen replaced the egg shells within nests, effectively simulating the behaviour of 'mutant' gulls that failed to remove shells. By doing this, he discovered that the white insides of the shells attracted predators to the nests and consequently the chicks had lower survival rates compared with nests where the shells were removed [46]. This was an elegant example of a field experiment in which the survival consequences of a specific behaviour were measured directly. That said, the results of experimental studies can be hard to interpret if the experimental manipulation of the IV falls outside the range of naturally occurring variation in this variable. In this sense, Tinbergen's experiment was a lucky exception.

Designing a good experiment involves a number of considerations including choosing appropriate controls, randomisation, blinding and choosing the optimal sample size, all of which we discuss in the following sections. These decisions are facilitated by online tools such as the Experimental Design Assistant (eda.nc3rs.org.uk; [47]).

4.4.1 Controls

In order to hold everything other than the IV constant, careful attention must be given to how the IV is manipulated and how subjects are allocated to different levels of the IV. In most experiments, it is difficult to vary one condition without varying something else as well. Part of the art of good experimental design lies in picking the appropriate control (comparison) groups and randomising or otherwise eliminating the effects of confounding variables.

The simplest experiments compare two groups of subjects, known as the **experimental** (or treatment) group and the **control** group. The effect on the DV (the so-called **treatment effect**) is measured in the experimental group and is compared with that in the control group, which is (at least in theory) similar in all other respects.

Suppose, for example, that psychologists wished to determine whether human cooperative behaviour is affected by exposing people to images of watching eyes, which might give them a sense of being observed. The experimental stimulus would feature an image of watching eyes, such as a poster with a full-face photo of a human face. Clearly, the control poster should not contain watching eyes. But what should it contain? A clever solution adopted in one study was to use a highly simplified representation of a face, comprising just three black dots arranged in a triangle (∵) with the two 'eyes' above the 'nose', as the experimental stimulus [48]. This is the simplest visual stimulus known to activate neurons in face-processing areas of the human brain. A control stimulus was created by simply inverting the image (∴) because the upside-down triangle of dots does not activate the face-processing neurons. This control was elegant because it was identical to the experimental stimulus in every respect except for the property the researchers wanted to manipulate.

4.4.2 Different Types of Experimental Design

The simplest experimental designs involve the manipulation of a single IV. Such experiments are relatively straightforward to conduct and analyse. However, they are limited to a single IV and do not permit inferences about other variables that might interact in important ways with the experimental manipulation. They would not indicate, for example, whether the manipulation was more effective in one sex than in the other. In more complex experimental designs, two or more IVs are varied simultaneously.

Factorial designs simultaneously measure the effects of two or more IVs, together with the interactions between these variables. Factorial designs are a more efficient alternative to conducting several simple experiments because they allow the effects of two or more IVs to be tested in the same group of subjects [49]. More importantly, factorial designs enable the detection of **interactions** between two or more IVs. An interaction means that the effects of two IVs do not simply add together but combine synergistically in some way so that a particular combination of treatments produces an effect that might not occur if the treatments had been varied one at a time.

Blocked designs control for known sources of random variation by allocating all treatment groups of the IV to all the different levels of a nuisance factor that is feared to affect the IV (the blocking factor). Suppose, for example, that researchers had conducted an experiment in two different rooms and were concerned that the conditions in the two rooms might be different. To control for this, they would ensure that all of the experimental treatments occurred in each room (known as 'blocking by room'). In this way, room would not be a confounding factor because it is not inadvertently correlated with one of the experimental groups. Moreover, if the block (the room, in this case) is included as a factor in the statistical analysis, comparisons can be made between treatments within each block, and the random variation arising from differences between blocks is effectively eliminated. Blocking can be done by, for example, space (e.g. field, room, cage), individual characteristics (e.g. age, sex, genetic strain) or time (e.g. time of day or season of testing).

A common form of blocking is the **matched pairs design**, in which the blocks are pairs of subjects that are likely to be more similar to one another than a randomly chosen pair of subjects, perhaps because they are twins, or come from the same nest, or have been matched by variables such as age, sex or socio-economic status.

The experimental designs described so far have involved comparing groups of subjects that are assigned to different experimental treatments. Such designs are known as **between-subjects designs** because each subject is only measured in one condition and the comparison is therefore between groups of subjects. The experimental equivalents of longitudinal correlational designs are **within-subjects designs**, in which each subject experiences different treatments sequentially, and measurements of the DV for the different treatments are compared within subjects. Each subject therefore acts as its own control, thereby eliminating the random variation due to differences between individuals (Box 4.1). Within-subjects designs are effectively a special case of blocking in which the block is the individual subject.

The advantage of within-subjects designs is the increased power resulting from the elimination of between-subject random variation. However, they

---

Box 4.1 Example of between- and within-subjects designs: vigilance in starlings

Consider the effect of the perceived risk of predation on vigilance in foraging starlings. Vigilance is measured as the number of times each minute a starling raises its head to scan for predators while it is foraging – the 'head-up frequency'. An experiment is performed to compare this behaviour in two treatments: one in which there are no cues of predator presence (the low-risk treatment) and one in which the birds are regularly exposed to cues of predators (the high-risk treatment). Head-up frequency is measured in 20 birds in the low-risk treatment and 20 birds in the high-risk treatment. A boxplot of the data shows a slight difference between treatments: the mean head-ups per minute score is 1.94 lower in the low-risk treatment, but this is not statistically significant (Figure 4.3a; two-sample t-test: $t_{38} = 0.78$, $p = 0.442$). However, when the experiment is repeated using a within-subjects design, in which each of 20 birds is measured twice, once in the low-risk treatment and once in the high-risk treatment, it becomes apparent that most birds reduce their vigilance in the low-risk treatment. The reduction is of the same magnitude (1.94 head-ups per minute), but this time the difference is statistically significant (Figure 4.3b; paired t-test: $t_{19} = 8.23$, $p < 0.001$). For the purposes of illustration, we have used exactly the same data points to create the two graphs and conduct the statistical analyses, which is why the mean difference between the high-risk and low-risk treatment is

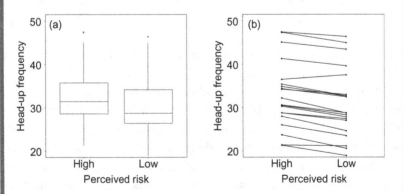

Figure 4.3 (a) Boxplot showing the distribution of head-up frequency for starlings in high- and low-risk treatments. (b) The individual data points used to create (a) but now connected by lines showing pairs of points coming from the same subject. All but two subjects have a lower head-up frequency in the low-risk treatment, despite the huge between-subject variation in each treatment.

> identical in (a) and (b). The only difference is the added assumption in
> (b) that each pair of data points comes from the same individual.
> In this example, the high random variation arising from large between-
> subject differences in the baseline level of vigilance obscures the
> treatment effect when a between-subjects design is used. When this
> source of random variation is eliminated by using a within-subjects
> design, the effect of treatment emerges clearly.

do have some disadvantages. As is the case for longitudinal correlational
studies, the repeated testing of the same subject may markedly influence its
behaviour. Processes such as arousal, sensitisation, conditioning, fatigue
and habituation may change the subject's behaviour, regardless of whatever
effect the experimental manipulation may be having. If these other changes
are ignored, the order of testing and the experimental treatment become
confounded, making the results uninterpretable. The point to remember is
that once an individual is tested, it becomes a somewhat different individual. Moreover, even if a subject's behaviour is not altered by the testing
itself, the subject will be slightly older at the second measurement and the
environment might have changed. These problems are usually addressed by
using a **counterbalanced** design, in which half of the subjects are presented
with the experimental treatment first, followed by the control treatment,
and the other half are presented with the experimental treatment second. If
more than two treatments are used, the order effects can be balanced
between different groups, with some individuals receiving one order of
presentation while others get another, so that every possible order is used.

4.5 Natural Experiments and Instrumental Variables

So far, we have presented correlational studies and experiments as though
they were strict alternatives. In fact, it is possible to obtain one of the main
benefits of an experimental manipulation – namely, the greater confidence
that the predictor variable causes the outcome of interest – without manipulating the predictor variable experimentally.

Natural experiments occur when an effectively randomly chosen sample of subjects experiences some kind of unintended influence or event, perhaps as a result of an environmental change or disaster. For example, researchers have assessed the effects of extreme maternal stress on behavioural outcomes for the baby by studying pregnant women who happened to be in the World Trade Center at the time of the 9/11 terrorist attacks in 2001 [50]. Events of this type have allowed scientists to study the effects of factors that could never be ethically simulated in a planned experiment.

When experimental manipulation of a predictor variable is not possible and has not occurred 'naturally', causality can be inferred using the method of **instrumental variables**. An instrumental variable is a variable that causes changes in the predictor variable but has no independent effect on the outcome variable. By substituting an instrumental variable for the predictor variable in an analysis, researchers can indirectly assess the likely causal effect of the predictor variable on the outcome variable in the absence of potential confounds associated with the predictor variable. For example, the price of alcohol affects how much alcohol pregnant mothers drink, but there is no reason to expect it to directly affect the baby's birth weight. The price of alcohol could therefore be used as an instrumental variable to test the causal effect of alcohol consumption (the predictor variable) on birthweight (the outcome variable).

An increasingly popular method called **Mendelian randomisation** uses common genetic polymorphisms (alternative genetic variants found in a population) as instrumental variables [51]. Consider again the question of whether smoking is associated with accelerated ageing. Researchers face the problem that smoking is correlated with many other factors, such as sex, poverty and education, which are also known to affect the rate of ageing. However, certain genetic polymorphisms have been identified that cause those individuals carrying them to have an increased likelihood of smoking. These polymorphisms have the advantage of being randomly allocated among individuals at the point of gamete formation (hence the name Mendelian randomisation), making them unlikely to be correlated with variables such as sex, poverty or education. Such polymorphisms can therefore be used as instrumental variables, in place of smoking behaviour, to determine with more confidence whether smoking causes accelerated ageing [52]. Of course, the Mendelian randomisation approach depends critically on the existence of genetic polymorphisms that are causally linked to the predictor variable of interest but have no direct effects on the outcome variable of interest, and such polymorphisms may not exist or be known. Large sample sizes are also required because the correlation between possessing a polymorphism and exhibiting a behaviour pattern is usually quite weak.

4.6 Randomisation

The purpose of correlational and experimental studies is to draw inferences about populations based on samples drawn from those populations. For these inferences to be valid, the sample studied must be representative of the population as a whole. Randomisation is the process of choosing a representative **sample** of subjects from the **population** of all possible subjects. If randomisation is done properly, every individual in the population has the same chance as every other individual of being a subject, and, in the case of experiments, every individual has the same chance of being assigned to each of the experimental groups. Randomisation is central to good study design because most statistical tests rely on an assumption that the sample was chosen randomly (see Chapter 11).

Perfect randomisation is hard to achieve. When studying humans, in particular, it is extremely difficult for researchers to obtain truly random samples because the subjects recruited are likely to be to some extent self-selected. The subset of people who see a recruitment advert on social media or answer an email or phone call is very unlikely to be random, and the smaller subset who actually complete a study is likely to be further biased by factors such as age, sex or employment status. Thus, polling randomly chosen Facebook accounts or approaching people randomly in the street is unlikely to yield a random sample of the population.

Fortunately, there are ways around this problem. In very large-scale human studies, researchers sometimes attempt to recruit all individuals who were born within a specific time period. For example, the UK's National Child Development Survey targeted all babies born during the week of 3–9 March 1958 in England, Scotland and Wales, and 98 per cent of these individuals were included in the initial survey. An alternative approach is to use sophisticated procedures such as **stratified random sampling** for choosing samples that are representative of the population.

For example, researchers attempting to recruit a representative sample of UK citizens might seek to include the same proportions of people of different sexes, ages and socio-economic groups as exist in the UK population.

In experimental design, allocating subjects randomly to treatment groups is essential to avoid unintentionally introducing confounds into the design. The gold standard of experimental design is the **randomised controlled trial (RCT)**, in which subjects are randomly assigned to the experimental and control groups. The consequences of failing to randomise correctly can be subtle. For instance, the order in which researchers catch mice from their home cage for subsequent behavioural testing is likely to be influenced by individual differences in the subjects' fearfulness. Consequently, if the first half of the mice removed from their cage are all allocated to the experimental group and the remainder to the control group, the experimental treatment is likely to be confounded with baseline individual differences in fearfulness.

Established methods exist for allocating subjects randomly to experimental and control groups. One approach is to start by giving every subject a number (say 1–24). The numbers are then written on separate pieces of paper and placed in a bag. Numbers are drawn blindly from the bag and the first 12 are allocated to the control treatment and the second 12 to the experimental treatment. The same could be achieved by using a computer to randomly allocate 12 of the numbers to each group.

A practical problem that is often encountered in small-scale psychology studies is how to allocate human subjects to the experimental and control groups when the subjects are recruited sequentially. Ideally, the allocation would be done randomly – for example, by tossing a coin or using a pre-generated list of random numbers to determine each subject's group. In practice, however, when the total number of subjects is small, this procedure can lead to unbalanced group sizes by chance. One solution is to allocate subjects alternately to the experimental and control groups as they are recruited. While it may be hard to see why this procedure should introduce any systematic differences between the groups, the samples are not strictly speaking random, and some journal referees may object on principle. Researchers must think very carefully to ensure that any systematic means of allocation does not introduce a subtle confound. A better solution might be to employ a **pseudorandom allocation**, whereby the treatment is chosen randomly from a sample constrained to contain exactly equal numbers of the two treatment groups – for example, by choosing without replacement from a bag containing 12 black and 12 white balls.

Small samples are less likely to be representative of the population even if they are selected at random, due to chance variation. For this reason, it is common in experimental studies to artificially **balance** the numbers of subjects according to factors that could affect the DV, such as age and sex. Researchers often ensure that equal numbers of males and females are included in the experimental and control groups.

Different issues apply to correlational studies when considering randomisation. Correlational studies make use of naturally occurring variation in predictor variables. But this does not necessarily mean that the sample used in a study has to be chosen randomly from the population. In some situations, it may be better to selectively sample subjects chosen from the extremes of the distribution of the predictor variable in order to maximise the size of the effect being measured. For example, in a study to assess the relationship between economic deprivation and impulsivity, it may make sense to select subjects from the top 10 per cent and bottom 10 per cent of neighbourhoods, based on deprivation, rather than use a sample consisting mainly of subjects from neighbourhoods closer to the median.

4.7 Blinding

Researchers inevitably have some expectations about the outcome of an experiment, even if they are not consciously aware of them. These expectations may bias their observations in the direction of a favoured hypothesis. The cumulative effect of many such minor biases may be a false-positive difference between the experimental and control groups. The effects of unconscious bias are often surprisingly large, and the only sure way to minimise them is for the researchers to be genuinely unaware of how the subjects have been treated until *after* the data has been collected. This procedure is referred to as **blinding**.

Further complications can arise because of the potential influences of the researchers on their animal or human subjects. If the researchers are not blind to the treatment when the measurements are made, they may unconsciously influence their subjects through subtle behavioural cues. A famous example of such unconscious influencing was the case of Clever Hans, a performing horse that seemed to be able to count. Only as a result of testing under blind conditions was it found that the horse was not able to count but was, in fact, responding to subtle cues unconsciously produced by its trainer. Researchers should be mindful of how they may unintentionally affect the subjects' behaviour and consequently bias scores in the expected direction.

If the subjects of a study are humans then they too may introduce bias into the results. If they are aware of the group they are in, or the treatment they have received, the subjects may form conscious or unconscious impressions about the expected results. In psychology, a **demand characteristic** is a cue that makes participants aware of what the experimenter expects to find or how participants are expected to behave. If at all possible, and ethically acceptable, human participants should be insulated from any demand characteristics by ensuring that they are not aware of which group they are in until after the study is over.

An experiment in which neither the people making the measurements nor the subjects know the treatment each subject has received is called a **double-blind** experiment. This type of experimental design is widely used in assessing the clinical effects of drugs and other forms of medical intervention where it is relatively easy to make the control and treatment manipulations perceptually indistinguishable – for example, by giving the control subjects realistic placebo pills or injections. In many psychology experiments, however, it is impossible for the participants to be blind to the manipulation they are receiving. An alternative tactic for shielding subjects from demand characteristics and reducing bias is to deliberately mislead them about the aims of the study. For example, in a study to discover the factors that predict opportunistic snacking, researchers would want to measure how many calories the participants consume when given the opportunity to snack. However, if the participants are aware of what is being measured, they might consciously or unconsciously alter their behaviour, perhaps by snacking less than they otherwise would. This problem could be addressed by telling the participants that the researchers are interested in measuring their snack food preferences rather than how much they eat (see Box 1.1). The ethical issues raised by deceit of this kind are discussed in Chapter 5.

4.8 When to Measure Behaviour

Choosing the right time of day at which to measure behaviour is an important practical issue in any study, whether correlational or experimental. Obviously, humans and animals are not equally active throughout the 24 h period, so the amount and type of activity seen will depend on the time of day at which the subjects are observed. Comparable arguments apply to seasonal variations in behaviour.

The problem of diurnal variations in behaviour can be approached in one of four ways. The first option is to record behaviour throughout the 24 h period, either by continuous observation or with several observation sessions spread across each day. This may not be feasible in many cases. A compromise might be to record at two or three times spread across each day. If the results obtained at the various times of day are markedly different, then they must be analysed and treated separately; if not, they can be pooled.

A second option is to record at a different time each day such that, averaged across the entire study, each part of the day is equally represented in the final sample. This approach does not work if the behaviour changes systematically from day to day – as, for example, when studying young, developing animals – or when behaviour undergoes marked seasonal changes.

A third option is to record at the same time each day. This is the most common approach, especially in laboratory studies. However, if all observations are made at the same time of day, the results should not be generalised to any other time of day. In practice, this limitation should not present great problems unless the time of day significantly influences the results. Problems can, however, arise when diurnal activity rhythms drift, or when comparing the behaviour of different populations or different species whose diurnal activity rhythms differ. Observations should, of course, be made at a time of day when the behaviour of interest is most likely to be occurring. A surprising number of researchers have studied the behaviour of nocturnal or crepuscular animals during the daytime.

The final option is to ignore the problem and record at a different time each day on a haphazard basis. This approach has no obvious merits, especially for experimental studies under laboratory conditions. Nonetheless, it is sometimes unavoidable when studying behaviour under difficult conditions in the natural environment. If the time of day that an observation is made is recorded, it may be possible to control for it in subsequent statistical analysis.

4.9 How Much Data to Collect?

As we saw in Chapter 2, many past behavioural studies have been under-powered. Collecting insufficient data makes false-negative results more likely and means that any positive results are more likely to be false positives. All else being equal, the more data the better, because statistical power is always improved by increasing the sample size (Box 4.2). At some point, however, there will be diminishing returns, when the marginal improvements in statistical power start to be outweighed by the resource and ethical costs of collecting further data. Deciding how much data is really necessary is therefore a critical part of study design. In behavioural research, this often boils down to deciding how many subjects to measure. The answer depends on other aspects of the design, such as whether the study is correlational or experimental, whether it is cross-sectional or longitudinal, and whether any experimental manipulations are performed within or between subjects.

4.9.1 Choosing the Right Number of Subjects

In most behavioural research, the unit of replication is the individual human or animal subject. Therefore, while there will be further decisions about how much data to collect on each subject, the first decision is how many subjects to measure. The answer depends on statistical power.

Statistical power is defined as the probability that a study will detect an effect when there is a true effect present to be detected (Box 2.1). In other words, power is the probability of obtaining a significant result (i.e. rejecting the null hypothesis) when the hypothesis under test is true. Formal **power analysis** is a method for calculating the optimal number of subjects [53].

Box 4.2 How statistical power changes with sample size and experimental design

Table 4.1 The statistical power for the two experiments shown in Box 4.1

	Between-subjects design	Within-subjects design
Statistical test	Two-sample t-test	Paired t-test
Number of birds (n)	20 per treatment group	20 in total
Effect size (difference in mean head-up rate between treatments)	1.94	1.94
Standard deviation (SD)	7.86	1.06 (SD of difference between treatments within birds)
Significance level (p)	0.05	0.05
Power	0.12 (very low)	1.00 (maximally high)

The example in Table 4.1 illustrates how a within-subjects design dramatically increases the power of an experiment. The between-subjects design is clearly underpowered and would most likely result in a false-negative error. In contrast, the within-subjects design is, if anything, *over*powered and uses more subjects than necessary. Neither experiment is optimally designed. By using formal power calculations, it is possible to plot how the power changes as a function of the number of subjects for each of the two designs (Figure 4.4). With a between-subjects design, more than 500 subjects (250 per treatment group) would be required to obtain an acceptable power of 0.8, whereas using a within-subjects design the total number of subjects required for a power of 0.8 is only 5. The plots also show how power is always a decelerating function of the number of subjects, so there comes a point where the increased power obtained by adding more subjects may not offset the extra costs involved.

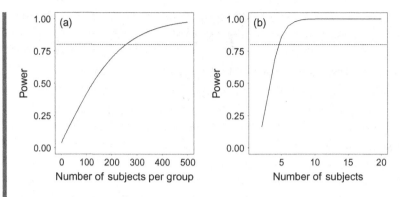

Figure 4.4 Power increases as the number of subjects per group increases.
(a) Between-subjects design, in which different subjects are measured in the
low-risk and high-risk treatments. (b) Within-subjects design, in which the same
subjects are measured in both treatments. The horizontal line shows a power of
0.8 in both graphs.

It involves specifying the statistical test that will be used to test the hypothesis and four out of the following five variables:

- The effect size to be detected (usually the difference between two means or a correlation coefficient).
- The random variation in the outcome variable (standard deviation).
- The acceptable risk of making a false-positive error, i.e. the significance level, which is conventionally set to 0.05.
- The acceptable risk of making a false-negative error, which is equal to 1 – power, where power is the probability of not making a false-negative error; by convention, a power of 0.8 is usually regarded as acceptable for an experiment, which means the risk of false-negative error is set at 0.2.
- The number of subjects.

Given estimates for any four of these variables, it is possible to calculate the fifth. Therefore, with estimates for the first four variables, it is possible to calculate the number of subjects required. Alternatively, if the number of subjects is fixed for some reason, it is possible to calculate the detectable effect size or the power achieved. Various online calculators and software packages will perform these calculations (e.g. InVivoStat: invivostat.co.uk; G*power: [54, 55]).

Statistical power therefore depends partly on biology because biology affects both the effect size and the amount of random variation present, and partly on decisions made by the researcher, who can determine the experimental design, the number of subjects and, to some extent, the measurement error.

While it is generally preferable to decide the number of subjects needed before starting to collect data, there can be exceptional circumstances in which **sequential sampling** could be justified – for example, if there were strong ethical reasons for keeping the number of subjects to the absolute minimum. In sequential sampling, analysis is repeated as the experiment proceeds and data collection stops when a predefined stopping criterion is reached [56]. This approach must be used with extreme caution to avoid accusations of *p*-hacking.

4.9.2 Choosing How Much Data to Collect on Each Subject

The amount of data that it is necessary to collect on each subject will depend on the within-subjects variation in the behaviour of interest. Some behaviours display little within-subjects variation – such as, for example, which hand a person chooses to write with. Such behaviours can be measured accurately with relatively little data, and it would be wasteful to measure each individual more than once or twice. However, most behaviours are less clear cut. When there is substantial within-subjects variation, random variation will be reduced and power increased by collecting more data per subject.

There will usually be a trade-off between measuring more subjects and measuring each subject more often. As a rule of thumb, unless the within-subject variation is very large, increasing the number of subjects will generally yield higher power than increasing the number of measurements of each subject.

4.10 Summary

- Study design is fundamental to good science. A poorly designed study will waste time and resources and could produce misleading or uninterpretable results.

- A good study design aims to minimise random variation and eliminate confounding variables.
- Correlational studies make use of natural variation in the variables of interest, whereas experimental studies manipulate variables to understand their causal effects on behaviour. There are advantages and disadvantages to both types of study, but experiments uniquely allow inferences about causation.
- A good experimental design requires subjects to be randomly allocated to experimental groups. Randomisation ensures the generalisability of results and eliminates confounds in experimental studies.
- Measurements should ideally be made blind to group membership. Blinding minimises biases caused by the conscious or unconscious expectations of the experimenter or subjects.
- Careful consideration should be given to when behaviour is measured, as time can affect behaviour.
- Power calculations can be used to determine the appropriate sample size.

5
Ethics and the Law

All researchers, including students, must think carefully about the ethical and legal implications of any behavioural research they are planning to conduct *before* they start. Research should be socially responsible. Public trust in science depends on scientists behaving legally and ethically. If this trust is undermined, the public are less likely to believe scientific findings, less likely to support the use of public money for scientific research and less likely to contribute as participants in research.

Research ethics refers to the set of norms that guide acceptable research practices. There are two central principles of ethical research. The principle of **research integrity** requires that research should be of sufficient quality to contribute to understanding and that it should meet recognised standards. The principle of **beneficence and non-maleficence** requires that the potential benefits of research should outweigh the risk of harm and that any such risks should be avoided or minimised. The benefits arising from research may be scientific, social or economic. Harms can affect the subjects of research or any parties with a stake in the research, including funders, institutions and society at large. Although much behavioural research appears relatively benign compared with some other areas of science, serious ethical issues relating to animal welfare and human rights can still arise.

Importantly, it is now widely recognised that stress in the subjects of behavioural research causes increased random variation in data [57]. More ethical research, in which more is done to ensure the well-being of subjects, will therefore go hand in hand with better-quality science.

Our aim in this chapter is to highlight the potential harms to animal and human subjects that can arise from behavioural research and show how these can be avoided or minimised.

5.1 Behavioural Research and the Law

Most countries have laws relating to the acceptable treatment of animals and humans, whether specifically in the context of research or more generally. Legislation relevant to research in animal behaviour encompasses laws that specifically govern the welfare of animals used in scientific research and legislation relating to the environment and to threatened and endangered species. Legislation relevant to research on human behaviour is diverse and includes laws governing the legal status and protection of children, the health and safety of workers, and the protection and use of personal data.

Nations differ substantially in their laws regarding research involving animals. For example, the UK's Animals (Scientific Procedures) Act 1986 (ASPA) protects all vertebrate animals and invertebrates of the cephalopod class, whereas the USA's federal Animal Welfare Act (AWA) covers only warm-blooded vertebrates and explicitly excludes birds, rats and mice bred for use in research.

Clearly, research must at the very least be lawful in the country in which it is conducted. It is the researcher's responsibility to find out about the legislation relating to research involving animal or human subjects in their country. Where research involves international collaboration between countries with different legal or ethical standards, the highest standards should be followed. This is the stated policy of many research funding agencies and scientific journals.

5.2 Formal Ethical Approval

Ethics and the law are related, but they are not always fully aligned. Just because something is lawful does not mean it is ethically acceptable. To ensure research integrity and help protect all those involved from unjustified harm, many institutions require proof that a formal **ethical review** has taken place and that approval has been granted. Proof of ethical approval is typically required by funders and universities *before* data collection can begin. Journals can refuse to publish research done without prior ethical approval. Failure to engage properly in the ethics process may result in unpublishable research, institutional sanctions and potentially, at worst, prosecution for unlawful behaviour. Ethical approval for a study cannot be obtained retrospectively.

Although applying for ethical approval can seem like a bureaucratic obstacle to scientific progress, it should be viewed positively as an opportunity to think hard about all aspects of a proposed project before leaping in to collect data.

5.2.1 Does All Behavioural Research Require Formal Ethical Approval?

As a general rule, all research involving human participants requires ethical approval. Most journals require evidence of formal ethical approval before they will publish data collected from human participants. The situation for research involving animals is more complex. The requirement to obtain ethical approval for an animal study, and the standards applied, will depend on a number of factors, including the species involved and the potential for harm to the subjects.

Animal research generally requires ethical approval if it involves **protected species**. Nations differ in how they define a protected species. As a general rule, vertebrates are likely to be protected and invertebrates are not, although there are important exceptions involving cephalopods and crustaceans [58]. Certain vertebrate species are afforded greater protection than others, based on their taxonomic proximity to humans (e.g. non-human primates), their rarity (e.g. endangered species) or their cultural significance (e.g. cats and dogs). No matter how harmless the research may seem, and whatever the local norms, prior ethical review should always be sought for research on vertebrate species or other protected species.

Research on non-protected species may still require ethical approval if it involves risk of harm to the environment. Emerging evidence on invertebrate cognition is fuelling a shift in thinking about the ethics of invertebrate use, with some commentators calling for ethical review of research on *all* invertebrate species [58].

Even if ethical approval for a study is not strictly required by national legislation or institutional rules, researchers are still advised to consult the *Guidelines for the Treatment of Animals in Behavioural Research and Teaching* [59], produced jointly by The Association for the Study of Animal Behaviour and the Animal Behaviour Society, or the *Guidelines for Psychologists Working with Animals* [60], produced by the British Psychological Society (BPS). If potential ethical issues are then identified, researchers should voluntarily seek ethical review.

Sometimes, a principal investigator or teacher will obtain ethical approval for a whole programme of related work. In such cases, it will not be necessary to obtain separate additional approval for each component study. Researchers or students working within larger programmes should make themselves aware of the ethical approvals in place and check that their studies are properly covered.

An important exception, where ethical approval may not be required, is research involving secondary analysis of pre-existing data. Such research does not normally require additional ethical approval, provided that the data was originally collected with any required approvals in place (and that this can be proven).

5.2.2 Ethical Review

The **research ethics committee (REC)** is an independent body responsible for reviewing and approving research proposals before research is conducted. Most universities and institutions where research is conducted have a REC. Research on some vulnerable populations, such as medical patients or prisoners, may require specialist ethical approval (e.g. from RECs in hospitals or prisons). It is common for institutions to have separate RECs for human and animal research, reflecting the differing expertise required. Independent researchers who lack an institutional REC should explore obtaining ethical review from a local university.

The typical ethical review process for both human and animal research is shown in Box 5.1.

5.2.3 When is Risk of Harm Ethically Justifiable?

Most ethical review processes adopt a utilitarian approach, whereby the risk of harm caused by a study is weighed against the potential benefits arising from the proposed research – a so-called **harm–benefit analysis** [61]. This means that research that risks harm can nonetheless be ethical if the

Box 5.1 The ethical review process

1. The researcher prepares an ethics application. RECs often have a standard form for this purpose. Depending on the level of risk involved in the proposed study, the details required are likely to include the number of subjects, protocols for the procedures involved, and copies of materials such as information sheets, consent forms and questionnaires. The details of the proposed study will therefore have to be decided before making an ethics application. It is common for RECs to include lay members who are non-scientists, and the proposal should be written for an educated non-expert, avoiding jargon. Many RECs explicitly request a **non-technical** or **lay summary**.

2. The researcher submits their application to their REC. Some RECs operate rolling ethical review processes, whereas others have periodic deadlines for committee meetings. It is worth discovering the local procedure and the length of time typically required to obtain approval.

3. The REC reviews the application and provides feedback to the researcher. The REC may have questions or request amendments that require a response from the researcher before approval can be granted. In the case of studies involving complex ethical issues, this can involve an iterative process that continues until the REC's concerns have been adequately addressed.

4. The researcher makes amendments to the application in response to the REC feedback.

5. The REC grants formal approval (or not) for the study. Approval may be subject to conditions; e.g. RECs often require periodic reports on progress and outcomes of research. RECs have the power to stop research if their conditions are breached.

6. The researcher can start data collection once the REC has granted formal approval.

7. The researcher should quote the name of the REC and the reference number for the formal ethical approval in any publications arising from the research. Many journals require this.

potential benefits are sufficiently high. For example, the Ethics Code of the American Psychological Association (APA) accepts that there are research topics of sufficient importance to society that the benefits of deceiving participants outweigh the harms caused by violating their rights [62].

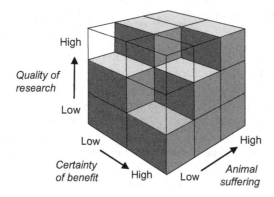

Figure 5.1 Bateson's cube, a tool for deciding whether animal research is ethically acceptable (the four clear blocks towards the front and top of the cube) or ethically unacceptable (the remaining solid blocks towards the back and bottom of the cube). Redrawn from [63] and [64].

Certain actions, however, are regarded as absolutely bad in themselves and their ethical acceptability is not affected by harm–benefit assessments [63]. For example, the AWA places an absolute ban on the use of paralytic drugs in unanaesthetised animals, whatever the potential benefits arising from the research. Similarly, the ASPA bans any invasive research on great apes.

In deciding whether a proposed study is ethically acceptable, at least three independent judgements must be made [63, 64]. First, what is the quality of the proposed research and how likely is it to advance scientific understanding? Second, how likely is the research to bring wider societal benefits – for example, to human and veterinary medicine, animal welfare, the economy or the environment? Third, how much harm to subjects or other stakeholders is likely to result from the research? These three judgements address the principles of research integrity, beneficence and non-maleficence.

The three dimensions of quality of research, certainty of benefit and likelihood of harm are brought together in the ethical decision cube shown in Figure 5.1. The cube, known as **Bateson's cube** after the ethologist Patrick Bateson, was originally developed for ethical decision making in animal research and the harm dimension is labelled 'Animal suffering' to reflect this. However, the cube could equally well be applied to human research with the suffering dimension relabelled 'Likelihood of harm'.

According to Bateson's cube, research of low quality is never ethically justifiable. Low-quality science will not bring scientific or societal benefits and could cause harm (see Chapter 12), which means that benefits cannot be used to justify any level of harm to subjects, however low the risks. The risk of even limited harm to subjects is ethically acceptable only when both the quality of the research and the certainty of societal benefit are judged to be high. High levels of harm are unacceptable, regardless of the quality of the research or its likely societal benefits.

Bateson's cube also says that research of high quality involving little or no harm is acceptable even if the work has no obvious societal benefits. This recognises the crucial long-term importance of understanding fundamental scientific phenomena, including those that have no immediate and obvious practical benefits. Such understanding is seen as a good in itself. History shows that it is seldom possible to predict exactly how understanding biology will in the future help to advance medicine or other fields, such as when ornithologists studying bird migration helped to understand the transmission of avian flu.

The three ethical dimensions embodied in Bateson's cube may seem hard to judge. Nonetheless, scientific quality is regularly assessed by committees distributing research funding and by editors of journals, and a consensus is usually reached when assigning research proposals to one of a few broad classes of merit. A similar argument applies to judging the likely societal benefits of a study; assessing the wider societal impact of proposed research is an integral part of many funding applications. For these reasons, RECs usually ask whether the proposed research has been subject to independent peer review – for example, in the form of endorsement by funding bodies – and take account of this information in their decision making. Judgements about potential harm require careful examination of what the research involves and the quality of the procedures to avoid or reduce any harm. Somewhat different considerations relate to research on animals and humans, and we cover these in the following sections.

5.3 Animal Research Ethics

Judging suffering in non-human animals is not straightforward. Different species evolved in different environments and have different senses and different biological priorities from those of humans. Consequently, species differ in what is harmful to them and in how they react to potentially harmful situations. For example, ultrasound emanating from an air conditioner is inaudible to humans but may be excruciating for rats. Judgements of animal suffering should be based on knowledge of the species' biology and natural history, not extrapolations from our own experience.

Although much animal behaviour research is non-invasive, this does not mean it cannot do harm. Keeping animals in unsuitable conditions can inflict considerable harm. Much animal behaviour research involves experimental manipulations that, while not conventionally invasive, can nevertheless inflict psychological suffering. Harm can result from denying animals pleasurable experiences, as well as from exposing them to unpleasant experiences. Field work historically received less ethical scrutiny than laboratory research, but there is now greater awareness that field studies can cause substantial harm [65]. Even research that only involves observation of animals in their natural habitats can potentially cause harm through physical disturbance of the environment, or through the researchers' effects on the predators or prey of the focal species. It is not just the treatment of animals during a study that is a source of potential harm. How animals are sourced, their general care before a study, the methods used for individually marking them, how they are handled by researchers, and their subsequent disposal or release at the end of a study are all potential sources of harm that should be carefully considered [59].

5.3.1 The 3Rs

The three principles of replacement, refinement and reduction, known as the **3Rs**, should be applied to all animal research in order to reduce animal suffering [66].

Replacement refers to avoiding the use of animals entirely. Although behavioural research usually involves the use of animals, computer simulations or video recordings can sometimes be substituted, especially in the context of teaching. **Partial replacement** involves replacing a species that is considered likely to suffer with a non-protected species considered far less likely to suffer based on current scientific evidence – for example, replacing a rodent species with *Drosophila* flies or nematode worms.

Refinement refers to changes to animal husbandry or scientific procedures that reduce potential suffering. If animals have to be transported, any sources of harm should be identified and minimised. Marking and tagging should be done using the least invasive methods. If animals are kept in captivity, then husbandry should be appropriate for the species' needs and adhere to published guidelines [67]. If a study necessarily involves a procedure likely to cause more than momentary pain or distress, appropriate relief measures such as anaesthesia, analgesia or sedation should be administered, unless this would clearly jeopardise the validity of the study. In long-term studies, behavioural training, whereby animals are conditioned to cooperate with procedures such as catching and blood sampling, should be used to reduce stress. Where animals need to be motivated to behave in a particular way, reward rather than punishment should be used wherever possible. Studies that involve one animal inflicting harm on another (e.g. studies of predation, aggression or intraspecific conflict) should wherever possible use field observations rather than staged encounters [59]. Where staged encounters are unavoidable, researchers should consider refinements such as the use of models or audio or video playbacks. Providing cover or escape routes for subjects could also reduce suffering. In research that is likely to involve considerable suffering, researchers should define and then apply humane end points (see section 5.3.2). Where euthanasia is necessary, it should be carried out as humanely as possible by trained personnel.

Reduction refers to reducing the number of individual animals used in a study to the minimum required to obtain acceptable scientific results. Reduction is achieved by careful preparation. Unnecessary replication of findings that are already well established can be eliminated by thorough literature searches. Inconclusive studies resulting from poor behavioural measures or ineffective manipulations can be avoided by small pilot studies. Using more animals than necessary, or too few to obtain significant results, can be avoided with power analyses. More data can be obtained from fewer animals through the use of good experimental designs paired with appropriate statistical analyses (see Chapter 4).

There may be occasions when the 3Rs conflict with one another. For example, it may be possible to reduce the total number of animals used in a study but only by increasing the degree or duration of suffering for the individuals that are used. In such cases, the researcher must strike a balance between reduction and refinement that minimises overall suffering without compromising the quality of science. The uncontrolled reuse of animals in sequential invasive procedures is ethically problematic because of the risk of cumulative adverse effects.

The UK's National Centre for the Replacement, Refinement and Reduction of Animals in Research (NC3Rs) supports the application of the 3Rs in animal research worldwide and provides many useful resources to assist researchers in this aim (nc3rs.org.uk).

5.3.2 Humane End Points

A **humane end point** is a predetermined early indicator or symptom that precedes more severe suffering. By terminating a painful or distressing procedure, giving treatment to relieve pain or humanely euthanising a subject when a humane end point is reached, suffering can be reduced while still meeting a study's objectives. The methods chosen for a study should avoid death as an end point because of the suffering likely to be experienced in the period before death. In laboratory studies of ageing or Darwinian fitness, where the longevity of animals is an outcome variable, natural death should be substituted with an earlier humane end point wherever feasible. See Box 5.2 for general guidelines on planning animal research.

5.4 Human Research Ethics

The BPS recommends that human subjects should be referred to as 'partici-pants', in recognition of their autonomy over their decision to participate in research and their active role in contributing to research. Researchers must respect the dignity and rights of human participants. The *Code of Human Research Ethics* of the BPS [70] sets out general principles intended to cover all research with human participants. Other national societies, such as the APA, have published similar documents [62]. Although these documents contain much good advice, no code of ethics can cover all possible harms, and researchers should always apply their own judgement.

The BPS recommends that the risk of harm from research should normally be no greater than that encountered in ordinary life. Low-risk research of this type requires only light-touch ethical review. However,

Box 5.2 Planning guidelines for animal research

PREPARE [68], which stands for **P**lanning **R**esearch and **E**xperimental **P**rocedures on **A**nimals: **R**ecommendations for **E**xcellence, provides a detailed planning checklist relevant to all types of animal research including field studies. The rationale underlying PREPARE is that planning guidelines can assist funders, regulators and ethical review committees in assessing applications for research involving animals. The PREPARE checklist [69] for planning animal studies covers the following areas:

(a) **Formulation of the study**
1. Literature searches: systematic reviews, choice of species, hypotheses.
2. Legal issues: legislation for animal research, animal transport, occupational health and safety.
3. Ethical issues, harm–benefit assessment and humane end points: lay summary, 3Rs.
4. Experimental design and statistical analysis: animal numbers, randomisation, bias.

(b) **Dialogue between scientists and the animal facility**
5. Objectives and timescale, funding and division of labour.
6. Facility evaluation: physical suitability, staffing levels.
7. Education and training: staff competence.
8. Health risks, waste disposal and decontamination: risk assessment.

(c) **Quality control of the components in the study**
9. Test substances and procedures.
10. Experimental animals.
11. Quarantine and health monitoring: animal health and consequence for personnel.
12. Housing and husbandry.
13. Experimental procedures: refinements.
14. Humane killing, release, reuse or rehoming: relevant legislation, emergency killing.
15. Necropsy (post-mortem examination).

Not all items in the checklist will be relevant to all studies and the checklist can be adapted for different types of study. 'Animal facility' could refer to a farm, zoo or field site.

Box 5.3 Checklist for research considered to involve more than minimal risk (modified from [70])

Research that involves any of the following:

- Vulnerable groups including children, persons lacking capacity, and those in a dependent or unequal relationship to the researcher.
- Potentially sensitive topics including, but not limited to, abuse, sexual behaviour, illegal behaviour and body weight.
- A significant element of deception in which participants are deliberately misled in some way.
- Taking part in a study without knowledge. For example, covert observation of people in non-public places.
- Records of personal or confidential information including, but not limited to, medical records or biological information.
- Access to potentially sensitive data through third parties such as employers or doctors.
- The potential to induce psychological stress, humiliation, pain or more than mild discomfort, including research that may lead to potentially distressing categorisation of the participant, either by researchers or the participant themselves (e.g. 'I am stupid', 'I am abnormal').
- Invasive or abnormal interventions including, but not limited to, administration of drugs, vigorous physical exercise and hypnosis.
- The potential for adverse effects on employment or social standing, e.g. discussion of an employer.
- The collection of human tissue, e.g. blood or saliva samples.
- Financial inducements other than reasonable expenses and compensation for time. The concern here is that vulnerable participants may be tempted into actions they later regret.
- Prolonged or repetitive testing.

many research questions cannot be answered without taking some risks. Research involving more than minimal risk can still be ethical, but it requires more rigorous ethical review. Box 5.3 lists the types of research that are considered to involve more than minimal risk to human participants or other stakeholders. Whether research is low or high risk, a set of standard considerations applies to all human research. These considerations are intended to respect the autonomy of participants and reduce or eliminate the risks of harm.

5.4.1 Informed Consent and Right to Withdraw

Researchers should obtain informed consent from participants wherever possible and ensure that the participants understand their right to withdraw from the research at any time without penalty. Where subjects are children, or adults lacking mental capacity, consent should be sought from parents or those with legal responsibility. As part of the consent process, participants should be provided with clear information about the study, any risks involved and the time commitment expected. Participants should be told how their data will be stored and used and informed about the potential benefits of the research.

Consent is not required for observational research conducted in public places where the subjects could reasonably expect to be observed by strangers, such as airports, shopping centres or railway stations [70]. The distinction between public and private places can become blurred if data is collected in virtual locations such as internet chat rooms. The distinct ethical challenges presented by internet-mediated research have led the BPS to produce specific guidance [71].

5.4.2 Deception of Participants

Deception of participants is considered by some scientists to be wrong. Participants acting under false information may do things they would not choose to do if they had known the truth. Moreover, deception could reduce the availability of future research participants by making people suspicious or cynical about research [72].

Despite these concerns, some forms of deception are regularly used in research (Box 5.4) [72]. There is some debate about what constitutes deception. Many scientists would draw a distinction between withholding some details relating to the specific hypothesis under test, which could be regarded as acceptable, and deliberately providing false information to participants, which could be regarded as unacceptable. As a general guide, if participants are likely to be angry or upset when the deception is revealed at debriefing, the deception is potentially harmful and requires careful consideration and explicit justification. Researchers are often able to achieve the aims of a study without using deception.

> **Box 5.4** Types of deception commonly used in psychological research
>
> - Lying to participants to induce stress (e.g. telling them something bad has happened).
> - Failing to fulfil promises of payment.
> - Providing false feedback about performance in the study.
> - Providing false feedback about other participants' performance in the study.
> - Use of confederates who appear to be normal participants but are actually following the researcher's instructions.
> - Failing to make participants aware that they are subjects of research.
> - Providing participants with false information about a study's main purpose.
> - Providing participants with false or incomplete information about a product.

Scientific and academic disciplines differ in their attitudes towards the acceptability of deception. For example, deception is common in psychology but generally regarded as unacceptable in economics research [73].

5.4.3 Privacy and Confidentiality

Human subjects have a right to privacy, which in many countries is underpinned by legislation, such as the General Data Protection Regulation (GDPR) of the European Union [74]. An important part of privacy is protecting the confidentiality of personal data. Breaches of confidentiality have the potential to cause serious harm, including identity theft, ostracism, and loss of health insurance or employment. Researchers should consider whether it is necessary to collect information that allows the identification of participants, even if doing so might be lawful. Data should not be collected unless it is explicitly required for a study, and it should be protected to an appropriate standard as prescribed, for example, in data protection or privacy legislation.

One approach to protecting confidentiality is to limit access to the final dataset to those who have a clear need to know. However, tensions can arise between the necessity to protect participants' privacy and increasing pressure to make research data publicly available.

An alternative approach is to attempt to anonymise the data. Perhaps the most obvious way of doing this is to remove all identifying information such as names, dates of birth and addresses from the dataset. Although names and contact details are required on consent forms, these forms can be securely stored in a separate location from the data file and linked to it only via a participant number (an example of **pseudonymisation**). Various more sophisticated methods have been developed for anonymising personal data [75]. However, it is exceedingly difficult to achieve true **anonymisation** whereby it is impossible to link an individual's identity to their data: many studies have shown that supposedly anonymised datasets can be de-anonymised by comparing them with other publicly available data. It often takes only two or three data points to be triangulated to identify the individual subject.

5.4.4 Debriefing

In studies that require informed consent, the participants should be debriefed after they have made their contribution. Any deception should be revealed and justified during debriefing. If the research has exposed a medical or psychological problem of which the participant was unaware, then relevant advice should be provided.

5.5 Summary

- Public trust in science depends on scientists behaving legally and ethically. Ethical science is also often better science.
- To be ethical, research must be of sufficient quality to further scientific understanding and its potential benefits should outweigh the risks of harm to subjects or other stakeholders.
- All research must also be lawful.

- Conducting a harm–benefit analysis is central to ensuring that ethical standards are maintained in research and is required for the majority of behavioural studies.
- Formal ethical approval must be obtained before starting to collect data.
- Research on animals should minimise animal suffering by following the 3Rs principles of replacement, reduction and refinement. Humane end points should be used to limit unnecessary suffering.
- Research on humans should respect the autonomy and rights of participants and will generally require informed consent, the right to withdraw and debriefing. Deception is potentially harmful and should only be used following careful consideration.

6
Defining Behavioural Metrics

Measuring behaviour means assigning numbers to observations of behaviour according to specified rules. Henceforth, we use the term **metric** to refer to a defined behavioural variable that is measured. It is often the case that numerous different metrics could be used to test a given hypothesis. There may be different ways of measuring the same behaviour, such as the frequency with which the behaviour occurs or the total time spent doing it. Which metric is most appropriate will depend on the research question and the practicalities of measurement. We reserve discussion of choosing the right metric for Chapter 10.

The predictions to be tested in a study should always refer to specific metrics. A lack of clarity over which metrics will be analysed increases researcher degrees of freedom and is consequently a recipe for generating false-positive results. It is therefore essential to be clear about exactly what will be measured and how this will be done *before* starting to collect data. Behavioural metrics that are poorly defined are likely to yield unreliable measurements, especially if more than one observer is involved in collecting the data (see Chapter 10). Poor definitions also make it difficult for other researchers to replicate a study.

In this chapter, we describe some of the common types of behavioural metric and discuss the process of defining metrics unambiguously.

6.1 Categorising Behaviour

Behaviour consists of a continuous stream of movements and events. Before any feature of behaviour can be measured, this stream must be divided into discrete units or categories to which numbers can be assigned in some way.

6.1.1 Identifying Categories

In many behavioural studies, the problem of how to categorise behaviour is solved by using standardised behavioural tests. Such tests are designed to elicit the specific behaviour of interest and, often, to restrict the range of possible responses. In many such tests, subjects are presented with a choice: whether to press a lever; whether to turn right or left in a Y-maze; how much of a standard plate of food to eat; what proportion of a financial endowment to give away; and so on. The outcome of such choices is generally easy to quantify. We discuss the pros and cons of standardised tests further in Chapter 7.

If, instead, behaviour is observed in freely behaving subjects, the categories to measure may be less obvious. Freely behaving subjects may perform a wider range of behaviours than in a test situation and the question arises as to how many different categories to measure. In general, the number of categories used should be sufficient only to answer the research question posed and no more. Inexperienced observers sometimes err towards trying to record too much. A given stream of behaviour could potentially be divided into a very large number of categories, so it is essential to be selective.

Categories of behaviour should be **independent** of one another; that is, two or more categories should not simply be different ways of measuring the same thing. For example, measuring both the proportion of time spent being active and the proportion of time spent being inactive would be redundant if these were the only two categories possible. It is also best to drop categories that are not relevant to the research question, or which seem inconsistent and difficult to identify reliably. The chances are that the fewer categories used in a study, the more reliably each will be measured (see Chapter 10).

6.1.2 Describing Categories

Behaviour can be described in a number of different ways. One simple distinction is between describing behaviour in terms of its structure or consequences.

The **structure** is the appearance, physical form or temporal patterning of the behaviour. The behaviour is described in terms of the subject's movements and posture. The **consequences** are the effects of the subject's behaviour on the environment, on other individuals or on itself. The behaviour may be described without reference to how the effects are achieved. Categories such as 'obtain food' or 'escape from predator' are described in terms of their consequences and can be scored irrespective of the movements used to achieve them. 'Run tip of bill along primary feather of wing' is a structural description, whereas 'preen' is a description by consequence.

Describing behaviour by its structure can sometimes generate unnecessary detail and place heavy demands on the observer's ability to make subtle distinctions between complex patterns of movement. Description by consequence is often a more economical approach. However, description by consequence implies that the consequence of the behaviour is known, which may not always be the case. For this reason, it is best to use neutral terms for labelling such categories, rather than labels that falsely imply knowledge of the biological function of the behaviour or the subject's internal state. For instance, if a category of vocalisation is named 'distress call', rather than given a neutral label such as 'peep', an observer might be tempted to include vocalisations that did not meet the stated criteria for the category but which were emitted when the animal was apparently distressed.

A third type of description is in terms of the individual's **spatial relation** to features of the environment or to other individuals. In this case, the subject's position or orientation relative to something or someone is the salient feature. The emphasis here is not on what the subject is doing but where or with whom. For example, categories such as 'approach' or 'leave' could be defined in terms of changes in the spatial relation between two individuals.

6.1.3 Defining Categories

In some cases, behaviour is composed of discrete units of clearly distinguishable and relatively stereotyped actions, such as pecks, grunts or items of food consumed, in which case the definition is largely dictated by the behaviour itself. More often, however, the definition of a category will depend on the question being asked, rather than some inherent feature of the behaviour. Take, for example, the definition of a 'meal': answering some research questions may require a meal to be defined as any consumption event, whereas other questions may require a definition that specifies a minimum number or amount of different foods consumed within a given time period.

Each category of behaviour to be measured should be clearly, comprehensively and unambiguously defined, using criteria that can easily be understood and applied by other observers. Good definitions specify clearly the actions required to determine whether the category of behaviour has occurred. For example, a bird could be defined as eating whenever the tip of its bill is below the rim of the food dish.

The criteria used to define a category should distinguish it from other categories, particularly those it resembles most closely. A detailed definition of each category and the associated recording method should be written down *before* starting to collect the data that will be used in the final analysis. This is essential to prevent definitions from shifting during the course of the study – a problem known as **observer drift**. All the data for a category used in the final analysis must, of course, be strictly comparable, and data obtained before the final definition was formulated should be discarded.

A period of **preliminary observation** provides an opportunity to refine the criteria used to define each category. A wholly satisfactory and unambiguous definition of a category can rarely be formulated without having watched the behaviour for some time. Early definitions are often unable to deal with unforeseen ambiguous examples of the behaviour that crop up during preliminary observations. Developing a set of precise and unambiguous category definitions can take time.

6.1.4 Events Versus States

Most categories of behaviour can be classified as either events or states.

Events are behaviours of relatively short duration and typically low variance in duration, such as discrete vocalisations or body movements (e.g. lever press, binary choice, sneeze, blink, yawn). The salient feature of events is their frequency of occurrence. The number of times a dog barks in 1 min would be a measure of the frequency of a behavioural event.

States are behaviours of relatively long duration and typically higher variance in duration, such as periods of sleep, foraging, social interaction or feeding. The salient feature of states is their duration (mean or total duration, or the proportion of time spent performing the activity). The total time a dog spends asleep over a 24 h period would be a measure of the duration of a state. The onset or termination of a behavioural state can itself be scored as an event and measured in terms of its frequency. (The term 'state' is also used in the behavioural literature to refer to a motivational state, such as hunger or thirst. The two should not be confused.)

6.1.5 Ethograms

When choosing behavioural categories, it can be helpful to refer to existing descriptions of the main types of behaviour that typify the species. In some cases, such information is available in the form of a published **ethogram**, which is a catalogue of descriptions of the discrete, species-typical behaviour patterns that form the basic behavioural repertoire of the species. Unfortunately, ethograms vary enormously in the number of categories included and the detail with which these are described. Moreover, ethograms are not available for most species. A further limitation of ethograms is that not all members of a species behave in the same supposedly species-typical way, which means that an ethogram derived from one sample of animals may not accurately reflect the behaviour of another sample of the same species. Box 6.1 shows an example of an ethogram.

6.2 Behavioural Metrics

Once the behavioural categories for a study have been identified and defined, the next step is to formulate the specific metrics that will determine how the various types of behaviour are measured in practice. In general, the SI system of units (Système International d'Unités) should be used for all scientific measurements (Box 6.2). When behaviour is observed in freely behaving subjects, the resulting measurements commonly yield four basic types of metric: latency, frequency, duration and intensity (Figure 6.1). We describe these metrics and their associated units below.

6.2.1 Latency

The **latency** of a behaviour is the time from some specified event, such as the beginning of the trial or the presentation of a stimulus, to the onset of

Box 6.1 An ethogram specifically for assessing the welfare of captive rhesus macaques housed in cages in a laboratory setting (modified from [76])

The descriptions and definitions in the ethogram in Table 6.1 are structural: they focus on the subject's posture and movements rather than the consequences of its behaviour. One reason for using structural descriptions in this context is that many of the behaviours listed can have both a normal presentation and an abnormal presentation, in which fulfilment of the normal function does not appear to be the motivation for the behaviour. For example, self-grooming, self-scratching and body shaking have a normal function in hygiene and comfort, but they can also be performed at times of motivational conflict as so-called displacement activities. Differentiating between the normal and abnormal contexts can be difficult, which makes it safer to use neutral structural descriptions.

Table 6.1 Excerpt from ethogram

General category	Behaviours	Definition
Inactive	Alert	Sitting/lying/standing stationary on any surface and looking at objects or individuals inside or outside the cage
	Not alert	Sitting/lying/standing stationary on any surface, with eyes open or closed, not looking at objects or individuals inside or outside the cage
	Hunched	As for not alert but sitting with head lower than the shoulders
Non-social behaviours	Self-grooming	Stroking, picking or otherwise manipulating own body surface
	Self-scratching	Scratching the skin vigorously with the nails
	Yawning	Opening the mouth widely, teeth exposed, lips retracted without vocalisation
	Body shake	Dog-like body shake of the whole body
	Eye rub	Rubbing the eye with a hand

Of the behaviours listed above, yawning and body shake are of short duration and could be designated as events, whereas the other behaviours can potentially have longer and more variable durations and are better designated as states.

Three of the non-social behaviours listed in this ethogram, namely self-grooming, self-scratching and body shaking, have been shown to increase in frequency following dosing with an anxiogenic drug and decrease with an anxiolytic drug, validating them as anxiety-related displacement activities in macaques [77]. Since the frequency of each of these three behaviours is relatively low, it would be justifiable to pool them at the analysis stage to provide a single composite metric of anxiety (see section 6.3) [78].

the first occurrence of the behaviour. **Reaction time** is a latency metric used in many standardised behavioural tests in which the subject is presented with a stimulus or choice of some type, and the length of time taken to respond is recorded. A related metric is the **lag** between one event and another, such as the interval between one animal performing an act and another performing the same act.

Latencies are measured in units of time. If, as is normally the case in standardised tests, the period of observation is limited, the behaviour pattern may not occur at all during some tests, making it impossible to assign a latency value. One solution in such cases is to assign a latency value equal to the maximum length of the observation period.

6.2.2 Frequency

The **frequency** of a behaviour is the number of occurrences of the behaviour per unit of time. Frequencies are expressed as a number per unit of time and are measured in reciprocal units of time (e.g. s^{-1}, min^{-1} or h^{-1}). If a rat presses a lever 60 times during a 30 min recording session, the frequency of lever pressing is $2 \, min^{-1}$.

In many standardised behavioural tasks, the relevant metric is not the number of occurrences of a behaviour per unit time (i.e. frequency), but the

Box 6.2 Units of measurement and conventions for presentation

The SI system comprises seven base units, each of which is independently defined, plus various derived units that are defined in terms of combinations of base units or other derived units. Table 6.2 shows a selection of the SI base units and derived units that are most relevant to measuring behaviour.

By convention under the SI system, each unit is represented by a standard unit symbol such as m, s or kg. Symbols may be multiplied or divided by numbers or other unit symbols (e.g. 3 m, 0.112 kg, 16.5 m s^{-2}). Unit symbols are algebraic symbols and follow the conventions of algebra. They are not abbreviations and should not be followed by a full stop (except at the end of a sentence) or an 's' (to denote plural). The names of units (e.g. metre, second, kilogram) are all spelt with a lower-case initial letter. Symbols for units named after a person start with an upper-case letter (e.g. J for joule, Pa for pascal). Standard prefixes (e.g. k, c, m) are used to denote units multiplied by various powers of ten (e.g. 1 cm = 0.01 m, 1 km = 1000 m, 1 ms = 0.001 s).

Some non-SI units are also commonly used in behavioural science. Units of time greater than a second are generally measured in minutes, hours or days, represented by min, h and d. Other common departures from the SI system include temperature in degrees Celsius (°C) and nutritional energy in calories (Cal).

Table 6.2 Common SI base units and derived units

Unit	Symbol	Quantity	Equivalent in base units
metre	m	length	base unit
kilogram	kg	mass	base unit
second	s	time	base unit
hertz	Hz	frequency	s^{-1}
newton	N	force, weight	$kg\ m\ s^{-2}$
joule	J	energy, work	N m
watt	W	power	$J\ s^{-1}$
pascal	Pa	pressure	$N\ m^{-2}$

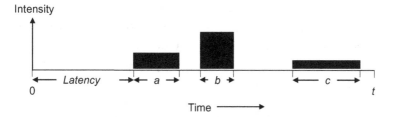

Figure 6.1 The meaning of latency, frequency, duration and intensity in a stream of behaviour. The black rectangles represent three successive occurrences of a behaviour during an observation period of length t units of time. Latency is the time from the beginning of the observation period to the first occurrence of the behaviour. Frequency is the total number of occurrences of the behaviour divided by the total observation time $(3/t)$. The total duration of the behaviour is $a + b + c$ units of time, and the mean duration is the total duration divided by three. Intensity is represented here by the height of the rectangles.

number of responses per opportunity to perform the behaviour, where opportunities are discrete 'trials'. For example, the number of times a subject chooses a particular outcome in a choice test would be expressed as a proportion (or percentage) of the total number of trials offered: if a rat chooses the left arm of a Y-maze five times out of a total of 20 trials, the frequency of left choices is 0.25 (or 25 per cent). Note that proportions and percentages are dimensionless indices with no units of measurement.

6.2.3 Duration

The **duration** of a behaviour is the length of time for which a single occurrence of the behaviour lasts. Duration is measured in units of time (typically s, min or h). If a kitten starts suckling and stops 5 min later, the duration of that period of suckling was 5 min.

'Duration' is also used in at least two other senses in the behavioural literature. The first is when duration (or total duration) refers to the total length of time for which all occurrences of the behaviour lasted over some specified period, usually the whole observation session. A total duration is, of course, meaningless unless the total time for which the behaviour was watched is also specified. To state that the total duration of a behaviour was, say, 16 min is meaningless: was it 16 min out of 20 min, an hour or a week?

To avoid ambiguity, a total duration should be expressed as the total duration over the specified period of observation (e.g. '9 min per 30 min') and should be explicitly referred to as **total duration**.

Alternatively, a total duration can be expressed as a proportion (or percentage) of the observation period, in which case it should be explicitly referred to as the **proportion** (or percentage) of time spent performing the behaviour. For example, if a kitten spent a total of 10 min suckling during a 30 min observation session, the proportion of time spent suckling would be 0.33. Expressing a duration as a proportion or percentage of total time omits the potentially important information about the total time for which the behaviour was observed. The confidence that could be placed in a statement that, say, the proportion of time a subject spent suckling was 0.33 would obviously depend on whether this figure was derived from several 24 h periods of observation or one 30 min observation. A **time budget** describes the proportions of time that a subject spends performing different categories of behaviour. Time budgets are usually expressed over a 24 h period.

Duration (or mean duration) is also used to refer to the mean length of a single occurrence of the behaviour pattern, measured in units of time. Mean duration is obtained by recording the duration of each occurrence of the behaviour and calculating the mean of these durations. To avoid ambiguity, this measure should be explicitly referred to as a **mean duration**. A mean duration can also be calculated by dividing the total duration of the behaviour by the total number of occurrences. This has the advantage that the duration of each occurrence need not be recorded separately. According to one definition of the term 'bout', the mean duration of a behaviour is equivalent to its **mean bout length.**

As an illustration of these various metrics, suppose that a mother and infant are observed for a period of 60 min, during which suckling occurred five times, with the individual periods of suckling lasting 3 min, 10 min, 1 min, 1 min and 1 min, respectively. According to the definitions given above, the durations of suckling were 3 min, 10 min, 1 min, 1 min and 1 min; the total duration of suckling was 16 min per 60 min; the proportion of time spent suckling was 0.27 (= 16/60); and the mean duration of suckling was 3.2 min (= 16/5).

Frequency and duration metrics provide different and complementary pictures. For instance, how often two monkeys groom each other (frequency) reveals something different about the nature of their social relationship from how long they spend doing it (duration). Frequency reflects the initiation of grooming, whereas duration reflects its continuation. Empirical studies have shown that frequency and duration measures of the same behaviour are not always highly correlated. It is therefore important to consider how the predictions being tested relate to the frequency and duration of a behaviour before starting to measure it.

6.2.4 Intensity

The term 'intensity' has no universal definition. It is generally used to describe the strength or amplitude of a behaviour, such as the speed or strength of an action, the loudness of a vocalisation or the brightness of a visual signal.

One simple metric of intensity is **local rate**. The local rate of an activity such as eating, walking or grooming is defined as the number of component acts per unit time spent performing the activity (Figure 6.2). Consider, for example, eating behaviour, which is composed of discrete, component acts of ingestion of individual food items. The local rate of eating would be given by the number of items ingested per unit time spent eating. Similarly, the intensity of walking could be measured by the number of strides per unit time spent walking, and the intensity of grooming by the number of stroking movements per unit time spent grooming. The local rate captures the speeded-up or hurried nature of intense behaviour.

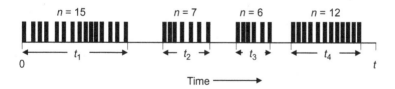

Figure 6.2 The meaning of local rate. Local rate provides a measure of the intensity of behaviour. An activity such as eating may be composed of discrete component acts such as ingesting a morsel of food, each act indicated by a vertical bar. The local rate is given by the total number of occurrences of the component act (in this case, the number of food morsels eaten during the observation period = 40) divided by the total duration of the activity (= $t_1 + t_2 + t_3 + t_4$).

Various direct metrics of intensity can be derived from behaviour with the help of technology. The intensity of some physical activities can be measured in terms of energy expenditure per unit time, the loudness of a vocalisation in decibels or heart rate in beats per minute. In some cases, intensity can be measured using a physical consequence of the behaviour as a proxy, such as the mass of food eaten per unit of time spent eating, the volume of water drunk per unit of time spent drinking or the distance travelled per unit of time spent walking. Intensity can also be assessed using subjective rating scales.

6.3 Composite Metrics

Two or more mutually exclusive behavioural metrics may be thought to represent alternative expressions of the same underlying propensity, or **latent variable**. If so, it can be informative to combine them into a single composite metric. For example, an animal presented with a threatening stimulus may react in a number of different and mutually exclusive ways, such as attacking, freezing or fleeing. If good grounds exist for supposing that all three of these responses indicate the same underlying motivational state (namely, fear), then pooling these metrics could make biological sense, even though they do not overtly measure the same behaviour. Creating composite metrics in this way has the advantage of reducing the number of different outcome variables that need to be analysed, which in turn reduces researcher degrees of freedom and the potential for false-positive errors.

Composite metrics can be created using either a top-down approach, which starts with a specific latent variable, or a bottom-up approach, which uses correlational analysis of the data to infer potential latent variables. A top-down approach would involve measuring the suite of behavioural changes caused by experimentally manipulating a latent variable. For example, anxiety-related behaviours may be defined as behaviours that increase in frequency when an animal has experienced potential threats in its environment (see example in Box 6.1).

Once such a group of related behaviours has been identified, the individual metrics should be standardised so that they have the same mean and variation before they are combined. This is usually done by calculating **z-scores** (the score for that individual minus the mean score for the sample, divided by the standard deviation). Measurements that are standardised in this way have a mean of zero and a standard deviation of 1.0. The composite metric for an individual is the sum (or mean) of the z-scores of the separate metrics. This procedure gives the same statistical weight to each metric, although different weights could be applied if such a procedure could be properly justified.

A bottom-up approach to creating a composite metric would use multivariate statistical tools, such as **principal components analysis (PCA)**, to create one or more composite factors (or principal components) that relate to different latent variables. Each statistical factor captures a certain amount of the total variance in the component metrics. In PCA, the maximum number of factors that can be derived is equal to the number of metrics. Ideally, a small number of factors would explain most of the variation in the data.

The success of statistical approaches such as PCA in creating useful composite metrics depends on how the various component metrics are correlated with one another. PCA will be of no use in creating composite metrics if the component metrics are completely uncorrelated. In contrast, the top-down approach is not dependent on any correlation between the component metrics within subjects. The top-down approach will work even if different subjects respond to the same latent variable in different ways – for example, if some individuals respond to fear by attacking, some by freezing and some by fleeing.

6.4 Levels of Measurement

When considering which behavioural metrics to use, a fundamental decision concerns the level of measurement required. Four different levels of measurement can be distinguished. In ascending order of strength of measurement, these are: nominal, ordinal, interval and ratio (Box 6.3) [79]. The level of measurement has important implications for data analysis and interpretation. Latencies, frequencies, durations and local rates are all measured on ratio scales.

Difficulties can arise when behaviour is measured using subjective **rating scales** or visual analogue scales (VAS) rather than objective measurements. A rating scale is a set of discrete options designed to elicit information from

---

Box 6.3 Definitions of levels of measurement

Nominal. If observations are simply assigned to mutually exclusive qualitative classes or categories (e.g. moving/not moving, active sleep/ quiet sleep/awake, type A/type B/type C/type D), then the variable is measured on a nominal (or **categorical**) scale. If only two outcomes are possible (e.g. yes/no, right turn/left turn), then the variable is said to be **binary**.

Ordinal. If the observations can also be arranged along a scale according to some common property (i.e. A > B > C >...) then the variable is measured on an ordinal (or ranking) scale. When a set of measurements is arranged in this way, they are said to be **ranked**. The number assigned to each measurement is its **rank**, while the arrangement as a whole is called a **ranking**.

Interval. If, in addition to being ranked, the scores can be placed on a scale such that the distance between two points on the scale is meaningful (i.e. the *difference* between two scores can be quantified), then the variable is measured on an interval scale. The zero point and unit of measurement are arbitrary for an interval scale. A temperature measured in degrees Celsius is measured on an interval scale.

Ratio. The highest level of measurement is attained when the scale has all the properties of an interval scale and also has a true zero point. This is referred to as a ratio scale because, in contrast to interval scales, ratios of measurements are also meaningful. Having a non-arbitrary zero point makes it meaningful to state, for example, that one animal sleeps for twice as long as another or moves half the distance. Mass, length and time are measured on ratio scales.

---

an observer about a quantitative or qualitative attribute. For instance, a rating scale for estimating the distance between a primate mother and her infant in a natural environment could comprise the following possible scores: 0–2 m, 2–4 m and >4 m. An equivalent VAS would require the observer to estimate the distance by indicating a position along a continuous line with designated end points (e.g. 0 and 4+ m). The subjective pain experienced by humans is generally scored on a VAS; one commonly used scale asks the person to rate their pain on a scale ranging from 0 (no pain) to 10 (the worst pain imaginable).

Scores on such scales can be converted to numbers for the purposes of statistical analysis. However, assigning numbers does not necessarily mean that the behaviour is measured on an interval or ratio scale. For example, subjectively rating pain on a scale of 0–10 would not constitute measurement on a true interval scale because there is no basis for assuming that the difference between scores of, say, 1 and 2 is the same as the difference between scores of 7 and 8. Such scores can be ranked, but the *differences* between scores may not be meaningful and therefore the measurement would be on an ordinal scale.

Even when measurements are made on a true ratio scale, it may not always make biological sense to interpret the results as ratio-scale measurements – as, for example, in the measurement of colour. While the reflectance spectrum of an object can be measured objectively using a spectrophotometer, the subjective impression the object has on an animal or person will depend on their eyes and brain because species and individuals differ in the frequencies of light their eyes can detect.

6.5 Summary

- Measuring behaviour means assigning numbers to observations of behaviour according to specified rules.
- Converting a stream of behaviour into behavioural metrics involves choosing and defining specific categories of behaviour that can be measured.
- Behavioural categories can be described in terms of their physical structure or their consequences.
- An ethogram is a catalogue of the species-typical behavioural categories displayed by a species in a specified environment.
- Descriptions of behavioural categories should be unambiguous and written down before data collection starts.
- Behavioural categories can be designated as either events (short duration) or states (longer duration).
- Behavioural categories are used to generate metrics such as latencies, frequencies, durations and intensities.
- Two or more metrics can be combined to form a composite metric.
- Metrics can be at different levels of measurement, ranging from nominal (weakest) to ratio (strongest).

7
Recording Methods

Having settled on the behavioural metrics required to answer a research question, the next step is to decide on the recording method – that is, how the behaviour will be recorded in practice. Two basic decisions must be made about the recording method: first, whether to record behaviour in the laboratory or in the natural environment ('the field'), and second, whether to use a standardised behavioural test or observe spontaneous behaviour. In many cases, similar metrics can be obtained using different methods, but each has its own advantages and disadvantages.

In this chapter, we focus on the general principles of recording behaviour, which are largely independent of the physical recording medium used to capture the measurements; the technologies used for that are considered in Chapter 8.

7.1 Laboratory Versus Field

Behavioural studies can be conducted in the laboratory or the field. We use 'laboratory' here in a broad sense to describe any situation in which behaviour is recorded in selected subjects that are contained, however briefly, in a standard environment controlled by the researcher. In contrast, we use 'field' to describe any situation in which behaviour is recorded in subjects that can behave freely in their home environment and interact with their own and potentially other species. Field studies need not necessarily be carried out on wild animals; the conditions on a farm or in a large zoo enclosure might have more in common with a field study than a laboratory. Field studies of human behaviour may take place in environments such as city streets, schools and workplaces. In practice, there is a continuum between laboratory and field conditions.

Conducting research in a laboratory setting has obvious advantages. The sample of subjects studied can be selected to yield individuals with known characteristics and histories (e.g. sex, age, genotype, health, education). The conditions in a laboratory can be controlled more easily, thereby eliminating many sources of random variation in behaviour. Recording methods can also be more standardised in the laboratory.

Laboratory studies have disadvantages, however. A captive subject is usually too constrained by its artificial environment to perform even a fraction of its normal behavioural repertoire. Humans and other animals often behave differently when they know they are being observed (so-called **observer effects**). These unwanted influences are very difficult to avoid in the laboratory, especially in the case of human participants. Furthermore, evidence that a particular factor influences behaviour in the laboratory does not necessarily mean that it influences the behaviour of freely behaving individuals in a similar way. Laboratory studies are more likely to yield reliable data, but the extent to which the results generalise to other environments may be questionable.

To observe the full richness of behaviour, subjects must be studied in the field. Fieldwork can uncover previously unrecorded aspects of behaviour, providing raw material from which new hypotheses may be formulated. Moreover, fieldwork provides an understanding of how a species' behaviour is adapted to the social and ecological conditions in which it normally lives.

The practical difficulties inherent in fieldwork should not be underestimated. The conditions for recording behaviour in the field are rarely ideal, and it may be difficult to collect high-quality data. A subject under observation may disappear from view, spoiling the best-laid plans for systematic recording. Ensuring randomness or balance in the choice of subjects may not be easy because some individuals are difficult to find. Bad weather may make observation impossible. Animals may be shyer than expected and require months or even years of habituation before they allow researchers close enough to make effective observations. Humans are often difficult to observe undetected, even when this is ethically acceptable. Furthermore, populations may differ in their behaviour, so just because a study has been conducted in the field does not mean that the results from one population will necessarily generalise to another.

In most cases, there will be a trade-off between the more controlled conditions of the laboratory and the greater ecological validity of the natural environment. Which is more important will depend on the question being asked.

7.1.1 Where Does the Internet Fit In?

The internet is widely used for behavioural research on humans because it enables huge datasets to be collected relatively easily [80]. Internet studies can to some extent mimic the conditions of either laboratory or field research as defined above.

In laboratory-type internet research, participants are recruited via social media or specialised sites (e.g. prolific.co) and take part in online questionnaires or behavioural tasks. Such studies have some of the qualities of laboratory research, in that the subjects know they are participating in a study and the procedures are standardised. Recruitment over the internet also makes it more feasible to target specific subsets of subjects, chosen according to characteristics such as age, sex, wealth, health, lifestyle, location or beliefs. Even so, such studies lack much of the control that is possible in carefully designed laboratory research. Researchers have little or no control over the environment in which participants take the test: they may be alone in their bedroom at home, in the park or on a crowded train. The relatively poor quality of the data can be offset to some extent by using a much larger sample size than would be feasible in traditional laboratory research. Nonetheless, the lack of control over certain variables may be critical in some cases, such as studies in which the behaviour being measured is sensitive to the number of people around the subject at the time of testing. Failure to control or record such variables could result in false-positive or false-negative errors, no matter how large the sample size.

In field-type internet research, the online behaviour of subjects is observed covertly by making use of publicly available information, such as the content of online conversations in internet chat rooms or the structure of social networks on social media sites. In such cases, researchers will have to demonstrate that their research is compliant with ethical standards and privacy legislation.

7.2 Standardised Behavioural Tests

In many areas of behavioural science and psychology, researchers have developed behavioural tests that are designed to reliably elicit specific behaviours under standardised conditions. By their nature, standardised tests tend to constrain the possible behavioural responses of the subject, thereby reducing the complexity and variability of the behaviour to be measured. This is both a strength and a weakness.

Standardised tests usually consist of discrete **trials**, each of which has a clearly defined beginning and end, thereby removing ambiguity over when to start and stop measuring behaviour. They typically also have clearly defined behavioural metrics associated with them. Two further advantages of established tests are that they are likely to have been validated and there will be existing data against which new data can be compared.

Most standardised tests are designed for laboratory research in which the subjects are introduced into the test situation by the researcher. Box 7.1 gives examples of common laboratory tests for measuring anxiety-like behaviour in mice and rats. Standardised tests have also been devised for the field, such as observing the reactions of animals to playbacks of conspecific alarm calls or a predator model, or observing the reactions of members of the public to a leaflet placed on their car windscreen or to an accomplice dropping a bag of shopping in the street.

Standardised tests are not without problems [81]. Unintended variations in methodology between laboratories could lead to substantial variation in the behaviour measured. One major problem is the lack of consensus on the precise behavioural metric used, raising concerns over increased researcher degrees of freedom and potential for *p*-hacking (see Chapter 2). A further risk arising from the widespread adoption of a small number of standardised tests is that the resulting scientific conclusions may be built on shaky foundations if these tests are not in fact measuring the constructs they are supposed to measure [82].

Unsurprisingly, standardised tests do not exist for many types of behaviour in many species. Before embarking on designing a new test, it is worth searching the literature to see whether suitable tests already exist for the behaviour of interest in a different species. It may then be possible to adapt an existing test. If so, this should be done with care. The development of a good standardised test relies on a sound understanding of a species' normal behaviour and ecology, and mistakes can be made if this understanding is lacking. For example, standardised tests of anxiety in rats and mice rely on the assumption that they avoid open, brightly lit or elevated spaces when they perceive potential threats in their environment. The elevated plus maze, the open field and the light/dark box (Box 7.1) all capitalise on this

Box 7.1 Standardised behavioural tests for anxiety in rats and mice

The four standardised behavioural tests in Table 7.1 are used extensively with rats and mice in laboratory studies of the neurobiological basis of anxiety and in screening for novel anxiolytic drugs. In all cases, the subjects are released into the apparatus at the beginning of a trial and left to explore the apparatus for 5–10 min.

There is no single, universally agreed metric of anxiety for any of the above tests, and different researchers have reported different metrics [82]. The different metrics are rarely concordant within the same study, suggesting that they are measuring slightly different things.

Table 7.1 Examples of standardised tests

Test	Description	Anxiety metrics
Elevated plus maze (EPM)	The EPM comprises four arms in a plus-sign shape (+); it is elevated from the ground with two opposite walled arms and two opposite open arms. The subject is released in the central area.	Open arm entries; enclosed arm entries; difference between open and enclosed arm entries; percentage open arm entries; percentage open arm time.
Elevated zero maze (EZM)	The EZM comprises an elevated circular runway divided into two enclosed quadrants opposite two open quadrants. The subject is released in one of the enclosed quadrants.	Enclosed-to-open crossings; percentage time in open quadrants.
Light/ dark box (LDB)	The LDB comprises two chambers, one lit and the other dark, connected by a tunnel. The subject is placed in one of the chambers.	Lit-to-dark crossings; dark-to-lit crossings; time in lit chamber; percentage of time in lit chamber.
Open field (OF)	The OF comprises a cylindrical, rectangular or square box with an open top. The subject is released into the central area away from the walls.	Latency to leave central area; number of entrances to central area; total time in central area.

species-typical response. However, other species respond differently to threats. For example, passerine bird species such as the starling respond to threats either by freezing or by seeking out high perches that provide a good vantage point and aerial escape routes. Clearly, the rationale underlying rodent tests for anxiety would not apply to starlings and a completely different kind of test is required.

Standardised tests may also be unsuitable for practical or ethical reasons, in which case researchers must rely on observations of spontaneous behaviour.

7.3 Observation of Spontaneous Behaviour

Answering some questions will require researchers to observe the spontaneous behaviour of freely behaving subjects, possibly over prolonged periods or in complex social situations. Research relating to the function of behaviour will often need to be conducted in natural environments where animals face selection pressures such as food shortage and predation. Observing freely behaving subjects in their natural environments is also important for establishing the ecological validity of standardised tests.

Designing a study to measure the spontaneous behaviour of freely behaving subjects requires two basic decisions. The first, which we refer to as the **sampling rule**, specifies which subjects to watch. This covers the distinctions between ad libitum sampling, focal sampling, scan sampling and behaviour sampling. The second decision, which we refer to as the **recording rule**, specifies how the behaviour of the chosen subject is recorded. This covers the distinction between continuous recording and time sampling, which in turn is divided into instantaneous sampling and one–zero sampling. The hierarchy of sampling rules and recording rules is shown in Figure 7.1.

7.4 Sampling Rules

7.4.1 Ad Libitum Sampling

With ad libitum (or free) sampling, no systematic constraints are placed on what is recorded or when; the observer simply records whatever is visible

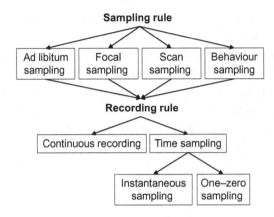

Figure 7.1 Hierarchy of sampling rules and recording rules. It is always necessary to specify both a sampling rule and a recording rule, although not all recording rules are appropriate for all sampling rules. For example, if scan sampling is chosen, then recording must be by instantaneous sampling.

and seems relevant at the time. An obvious problem with this method is that observations will be biased towards those behaviours and individuals that happen to be more conspicuous. Ad libitum sampling tends to miss brief behaviours and underestimate the involvement of some individuals in social interactions [83]. Provided this limitation is borne in mind, ad libitum sampling can be useful during preliminary observations or for recording rare events.

7.4.2 Focal Sampling

With focal sampling, one individual (or one dyad, one litter or some other unit) is observed for a specified amount of time. Ideally, the choice of focal individual is determined before the observation session, and the sequence in which focal individuals are watched is varied systematically. When recording the social behaviour of a focal individual, it may be necessary to record certain aspects of other individuals' behaviour, such as who initiates interactions and to whom the behaviour is directed.

At times, the focal individual may be partially obscured or move completely out of sight, in which case recording must stop until it is visible again. Any such interruption should be recorded as 'time out' and the final metrics calculated according to the time for which the focal individual was visible. Be aware, however, that omitting 'time out' can introduce bias if subjects systematically tend to do certain things while out of sight. For instance, many animals seek privacy when eating or mating, which means their behaviour when visible to the observer is not fully representative of their behaviour as a whole. Focal sampling can be difficult under field conditions because the focal individual may leave the area and disappear.

Some authors confusingly use 'focal sampling' as a synonym for continuous recording, conflating a sampling rule (who is watched) with a recording rule (how their behaviour is recorded). In fact, any of the three recording rules (continuous recording, instantaneous sampling or one–zero sampling) can be used when focal sampling the behaviour of a single subject.

7.4.3 Scan Sampling

With scan sampling, a whole group of subjects is rapidly scanned, or 'censused', at regular intervals and the behaviour of each individual at that instant is recorded. The behaviour of each individual scanned is, necessarily, recorded by instantaneous sampling. Scan sampling usually limits the observer to recording only one or a few simple categories of behaviour.

The time for which each individual is observed in a scan sample should, in theory, be negligible. In practice, when conducted by a human observer, it is at best short and roughly constant, with a single scan taking anything from a few seconds to several minutes, depending on the size of the group and the amount of information recorded for each individual. Automated methods can provide a closer approximation to true scan sampling.

Scan sampling may introduce bias if some individuals or some behaviours are more conspicuous than others. Focal sampling when the subjects are continuously visible is not subject to this bias.

Scan sampling can be used in addition to focal sampling during the same observation session. For instance, the behaviour of a focal individual could be recorded in detail, and at fixed intervals (e.g. every 10 or 20 min), the whole group could be scan sampled for a single category such as the predominant activity or proximity to each other.

Some authors confusingly use 'scan sampling' to refer to instantaneous sampling, again conflating a sampling rule with a recording rule.

7.4.4 Behaviour Sampling

With behaviour sampling, the observer watches a whole group of subjects and records each occurrence of a particular type of behaviour, together with information about which individuals were involved. Behaviour sampling is used mainly for recording relatively rare but significant types of behaviour, such as fights or copulations, where it is important to record each occurrence. Rare behaviours would tend to be missed by focal or scan sampling. Behaviour sampling is often used in conjunction with focal or scan sampling. It is subject to the same source of bias as scan sampling, in that conspicuous behaviours are more likely to be seen. Indeed, behaviour sampling is sometimes referred to as 'conspicuous behaviour recording'.

7.5 Recording Rules

Having decided which subjects to watch (the sampling rule), the next stage is to decide how to record their behaviour (the recording rule). Recording rules are of two basic types: **continuous recording**, which aims to provide an exact record of the behaviour by measuring true frequencies and durations of behavioural events and states, and **time sampling**, which involves sampling the behaviour periodically. Time sampling, in turn, is divided into two main types: **instantaneous sampling** and **one–zero sampling**.

Time sampling is a way of condensing information and thereby enabling the observer to record several different categories of behaviour simultaneously. In order to do this, the observation session is divided into successive short periods of time called **sample intervals** (Figure 7.2). The instant of

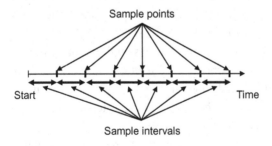

Figure 7.2 Explanation of sample point and sample interval.

time at the end of each sample interval is referred to as a **sample point**. For example, a 30 min observation session might be divided into 15 s sample intervals, giving 120 sample points.

7.5.1 Continuous Recording

With continuous (or all-occurrences) recording, the observer records each occurrence of the behaviour of interest, together with information about its time of occurrence (Box 7.2).

True continuous recording aims to produce an exact record of the behaviour, including the times at which each instance of the behaviour occurred (for events), or began and ended (for states). For both events and states, continuous recording generally produces true frequencies. It also produces true latencies and durations if an exact time base is used. However, systematic bias can arise if a measurement of duration or latency is terminated before the bout of behaviour ends, either because the recording session finishes or because the subject disappears from view. This is because the longer a bout of behaviour lasts, the more likely its duration will be underestimated by the termination of recording.

Continuous recording captures more information about a given category of behaviour than time sampling and should be used when true frequencies or durations need to be measured accurately. Continuous recording should also be used when the aim is to analyse sequences of behaviour. It does, however, have practical limitations because it is more demanding for the observer than time sampling. One consequence is that fewer categories can be recorded reliably using manual methods.

7.5.2 Instantaneous Sampling

With instantaneous sampling (also known as point sampling or fixed-interval time point sampling), the observation session is divided into short sample intervals. At the instant of each sample point, the observer records whether or not the behaviour pattern is occurring (Box 7.2).

The measurement obtained by instantaneous sampling is expressed as the proportion of all sample points at which the behaviour of interest was

Box 7.2 Comparison of recording rules

Table 7.2 shows the scores calculated from the data shown in Figure 7.3 obtained from the three recording rules: continuous, instantaneous and one–zero.

Instantaneous sampling gives a good approximation to the actual proportion of time spent performing the behaviour, as calculated from continuous recording, and accurately records four separate bouts of behaviour. In contrast, one–zero sampling considerably overestimates the proportion of time spent performing the behaviour and records only two separate bouts.

Table 7.2 Calculation of behavioural metrics

Recording rule	No. of bouts	Bouts min^{-1}	Total duration (min)	Mean bout duration (min)	Proportion of time
Continuous	4	4 ÷ 16 = 0.25	4.0 + 1.2 + 1.7 + 2.2 = 9.1	9.1 ÷ 4 = 2.28	0.57
Instantaneous	4	4 ÷ 16 = 0.25	4 + 1 + 2 + 2 = 9.0	9.0 ÷ 4 = 2.25	0.56
One–zero	2	2 ÷ 16 = 0.13	5 + 8 = 13	13 ÷ 2 = 6.5	0.81

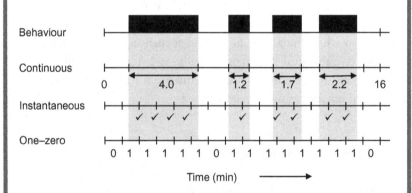

Figure 7.3 The black rectangles represent four successive bouts of a behaviour during an observation period lasting 16 min and divided into 1 min sample intervals. When using instantaneous sampling, occurrences of the behaviour are denoted with ticks on the sample point. When using one–zero sampling, any occurrence of the behaviour during the sample interval is denoted with a 1 (and an absence of the behaviour with a 0).

occurring. For example, if a 30 min recording session was divided into 15 s sample intervals, and the behaviour occurred at 40 out of the 120 sample points, the score would be 40/120 = 0.33. An instantaneous sampling score is a dimensionless index with no units of measurement.

Instantaneous sampling is best suited to recording behavioural states that can unequivocally be said to be occurring or not occurring at any point in time, such as measures of body posture, orientation, proximity, body contact or locomotor activity. Instantaneous sampling is not suitable for recording discrete events of short duration. Neither is it suitable for recording rare behaviours because a rare behaviour is unlikely to be occurring at the instant of any one sample point and therefore will usually be missed.

One potential source of bias with instantaneous sampling is the observer's natural tendency to record conspicuous behaviour patterns, even if they occur slightly before or slightly after the sample point. The sample point is thereby effectively extended from an instant to a window of finite duration, making the sampling no longer truly instantaneous. This is more likely to happen with conspicuous or significant behaviours, and consequently these behaviours will tend to be overestimated relative to less prominent behaviours.

Instantaneous sampling does not produce true frequencies or durations. Nonetheless, it can produce a record that approximates to continuous recording if the sample interval is short relative to the average duration of the behaviour. The shorter the sample interval, the more accurate instantaneous sampling is at estimating duration and the more closely it resembles continuous recording. Of course, using a very short sample interval largely negates the practical benefits of time sampling, in which case continuous recording might as well be used instead.

7.5.3 One–zero Sampling

With one–zero sampling (or fixed-interval time span sampling), the recording session is divided into multiple short sample intervals. At the instant of each sample point, the observer records whether or not the behaviour of interest has occurred at all during the preceding sample interval. The scoring is binary (yes/no) and is done regardless of how often, or for how long, the behaviour has occurred during that sample interval (Box 7.2). An equivalent procedure is to record the behaviour pattern when it first occurs, rather than waiting for the end of the sample interval.

The score produced by one–zero sampling is expressed as the proportion of all sample intervals during which the behaviour of interest occurred at least once. For example, if a behaviour occurred during 50 out of the 120 15 s sample intervals in a 30 min recording session, the score would be 50/120 = 0.42. As with instantaneous sampling, one–zero sampling produces a single dimensionless score for the whole recording session. The shorter the sample interval relative to the average duration of the behaviour, the more closely one–zero sampling approximates to instantaneous sampling.

One–zero sampling does not produce true or unbiased estimates of durations or frequencies. The proportion of sample intervals during which the behaviour occurred to any extent cannot be equated either with the length of time spent performing the behaviour or the number of times the behaviour occurred. This point requires even stronger emphasis than in the case of instantaneous sampling because one–zero sampling also introduces systematic bias. It consistently overestimates duration because the behaviour is recorded as though it occurred throughout the sample interval, even though that is often not the case. One–zero sampling also tends to underestimate the number of bouts because the behaviour could have occurred more than once during a sample interval. Both of these biases are illustrated in the example in Box 7.2.

Comparing one–zero scores, either between subjects or across different occasions, can be problematic unless the mean bout length of the behaviour remains roughly constant. This is because the error in estimating frequency or duration depends on the ratio of mean bout length to sample interval. Thus, if the mean bout length of the behaviour varies between individuals (or, for the same individual, varies between recording sessions), then the error in estimating frequency or duration will also vary.

The problems with one–zero sampling have led some authorities to argue that it should never be used. We do not agree. Our view is that one–zero sampling can be useful for recording complex and intermittent behaviour, such as play or certain types of social interaction, that start and stop repeatedly and rapidly, and last only briefly on each occasion. In such cases, continuous recording or instantaneous sampling may not be practicable because it is difficult to record each occurrence of the behaviour or specify at any one instant whether or not the behaviour is occurring, whereas it usually is possible to state unequivocally whether or not the behaviour has occurred during the preceding sample interval.

One–zero scores are valid measures of behaviour, in so far as they provide a meaningful index of the 'amount' of behaviour. One–zero scores are often highly correlated with frequency and duration measures of the same behaviour, which implies that they give a composite measure of the 'amount' of the behaviour. In contrast, frequency and duration measures of the same behaviour are not always highly correlated with one another. In some cases, therefore, a one–zero score may actually be a more meaningful index than either frequency or duration by itself.

7.5.4 Choosing the Sample Interval

The length of the sample interval used for time sampling should depend on how many different categories are being recorded, as well as the nature of the behaviour. The shorter the sample interval, the more accurate the record should be. However, the shorter the sample interval, the harder it is for the observer reliably to record several categories of behaviour at once, especially if the behaviour is complex or rapid.

In practice, observers must balance the potential accuracy of measurement, which requires the shortest possible sample interval, against ease and reliability of measurement, which requires an adequately long interval. If the sample interval is too short, observer errors may make the recording less reliable than if a slightly longer interval were chosen. The optimal sample interval is therefore the shortest possible interval that allows the observer to record reliably, after a reasonable amount of practice.

The optimal sample interval depends on what is being measured, and determining it may be a matter of trial and error. To give some idea, however, many researchers have used a sample interval in the range of 10 s to 1 min, with sample intervals of 15, 20 or 30 s being common under laboratory conditions. Field studies, especially those involving long recording sessions, may require longer sample intervals.

Instead of relying on trial and error, a systematic method could be used to determine the optimal sample interval, although this requires considerable additional work. First, a fairly large sample of the behaviour must be measured using continuous recording, in order to give an accurate picture of what actually happened. Scores are then calculated for each category as though the behaviour had been recorded using various different sample intervals (e.g. 10, 20, 30 and 40 s). The discrepancies between the continuous record and the simulated time sampling measures are then calculated for each sample interval. The error arising from time sampling will increase as the simulated sample interval becomes larger. It may be possible to distinguish a 'break-point', above which time sampling is unacceptably inaccurate and below which it yields a reasonable approximation to continuous recording. This point signifies the longest sample interval that can be used if the record is to be sufficiently accurate for that category of behaviour.

One obvious problem with this whole process, apart from the huge effort, is the need initially to measure the behaviour using continuous recording in order to provide an accurate record for comparison. In many cases, time sampling is used precisely because continuous recording is *not* practicable, which would obviously rule out this procedure. Another problem is that, in most studies, several different categories of behaviour are recorded, and the sample interval must be suitable for all of them.

7.5.5 The Pros and Cons of Time Sampling

A practical advantage of time sampling is that, by reducing the observer's work load, it enables more categories to be measured than would be possible with continuous recording. This can be an important consideration, especially in a preliminary study where more categories are recorded. Time sampling also enables the observer to study a larger number of subjects if individuals are watched cyclically, using scan sampling. For example, an observer could record the behaviour of a group of 12 subjects by watching them cyclically and recording the behaviour of each subject using instantaneous sampling every 15 s, thereby observing each subject once every 3 min.

Time sampling tends to be more reliable than continuous recording because it is less demanding. Moreover, some types of behaviour occur too rapidly for each occurrence to be recorded manually by a human observer, making time sampling a necessity.

The practical benefits of time sampling are achieved at the expense of capturing less information than is possible with continuous recording. Neither instantaneous sampling nor one–zero sampling produces accurate estimates of frequency or duration unless the sample interval is short relative to the average duration of the behaviour. Furthermore, neither is generally suitable for recording *sequences* of behaviour, unless the sample interval is very short. This is because with one–zero sampling, two or more instances of the behaviour can occur within the same sample period, while instantaneous sampling can miss changes in behaviour that occur between sample points.

Continuous recording and time sampling can be used simultaneously for recording different categories of behaviour within the same study.

7.6 Summary

- Behaviour can be recorded in either the laboratory or the field.
- In either setting, it can be recorded using standardised behavioural tests that elicit specific behaviour, or by observing freely behaving subjects.
- Observation requires decisions about which subjects to observe (sampling rules) and how to record their behaviour (recording rules).
- There are four sampling rules: ad libitum sampling, focal sampling, scan sampling and behaviour sampling.
- There are two basic types of recording rule: continuous recording and time sampling; the latter can be further divided into instantaneous sampling and one–zero sampling.
- Continuous recording is more demanding for the observer but is the only recording method that produces true frequencies and durations.
- Estimates of frequencies and durations derived from time sampling will be more accurate if the sample interval is short relative to the mean duration of the behaviour.
- One–zero sampling is likely to yield biased estimates of frequency and duration.

8
Recording Technology

Measuring behaviour involves transforming the rich, multimodal stream of actions of live animals or humans into the behavioural metrics needed to test predictions. In previous chapters, we discussed the general principles of this process, including how behavioural metrics are defined (Chapter 6) and the ways in which measurements can be elicited from subjects with standardised tests or sampled from spontaneous behaviour (Chapter 7). In this chapter, we outline the range of practical technologies that can be used to capture and process behavioural data.

8.1 The Technological Revolution

The nineteenth-century psychologist Francis Galton used the simple device of pricking a card concealed in his pocket with a pin to covertly record quantitative data on human fidgeting behaviour. While a researcher today is unlikely to use a method like this, it is nonetheless true that high-quality behavioural data can be collected using low-tech methods such as paper-and-pencil check sheets. The simplest technologies are still good enough to yield publishable data, but newer technologies can offer more.

The measurement of behaviour has been revolutionised by advances in technology, particularly in digital hardware and software. Some advances have enabled incremental improvements in data collection, such as the use of the internet to recruit participants for online studies, or the use of computer-based event recorders to increase the efficiency and precision of manually coding behavioural data. Other advances have been transformational, such as the deployment of miniature satellite-based global positioning tags on migrating birds to collect previously unavailable details of their flight paths, or the use of smartphones to poll large samples of people about their current behaviour in real time or to automatically track their movements and interactions with other smartphone users. The rapidly growing **internet of things** – electronic devices of diverse types connected to the internet – is creating ever more opportunities for remote automated collection of behavioural data from humans and animals in their normal environments.

New technologies have increased the volume of data it is possible to collect, the variety of data types that can simultaneously be measured and the velocity with which metrics can be extracted from raw data. The advent of so-called **big data** in behavioural science, characterised by the 3Vs of high volume, variety and velocity, is changing the types of questions it is feasible to tackle in animal and human behaviour. For example, the development of **algorithms** (sets of mathematical instructions or rules) that automatically extract behavioural metrics from video data makes it possible to conduct real-time automated monitoring of behavioural indicators as diverse as farm animal welfare and human criminal activity.

8.2 Data Pipelines

The ways in which technologies are used to record behavioural data are best understood in the context of data pipelines. A **data pipeline** describes the complete workflow, or sequence of processes, through which data is captured, recorded, processed and analysed, starting with the raw behaviour of live animals or humans and ending with the answer to a research question. The concept of a pipeline is borrowed from bioinformatics and is useful because it forces researchers to be transparent about their methods. In the past, critical steps such as the checking and 'cleaning' of data were often not reported, contributing to the replication crisis. The value of specifying a data pipeline becomes more obvious as workflows become more complex. A well-described data pipeline should allow an independent researcher to take the original raw data files and reproduce the results that appeared in a published paper.

At some point in the data pipeline, it is always necessary to **code** behaviour, by which we mean extract the required metrics from the complex stream of behaviour. Coding can be done either at the time of data capture or later, during data processing. Data capture and coding can be done manually, by the researcher, or automatically, using customised hardware or software. The data pipeline specifies the technologies and processes that are used at each stage.

Box 8.1 gives examples of three alternative pipelines for producing estimates of the proportion of time spent roosting by a caged bird. The

Box 8.1 Data pipelines for measuring behaviour

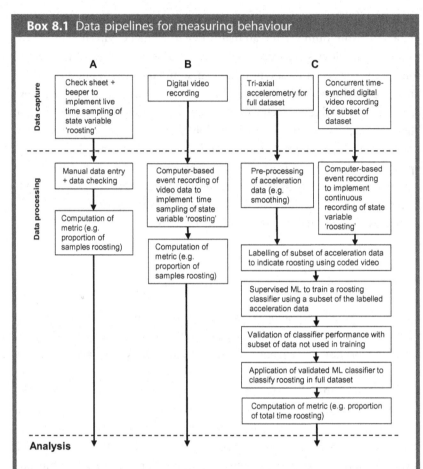

Figure 8.1 Examples of three alternative data pipelines for measuring the proportion of time spent roosting by a caged bird. Pipeline A shows a low-tech pipeline typical of classical ethology, B uses a modern ethological approach and C demonstrates a modern behavioural informatics approach using machine learning (ML).

Pipeline A shows one of the simplest data recording methods. The behaviour is observed live by the researcher and the technology for data capture is a paper check sheet used to implement time sampling with the aid of a device set to beep every 5 min. Roosting is defined as occurring when the subject is observed inactive on a high perch. Following data capture, the researcher manually enters the data into a computer for analysis. The main advantages are cheapness and simplicity.

Disadvantages include the loss of information due to time sampling (the state of the bird is only recorded at the sample points); the potential subjectivity of the primary data recording; and the inability to replicate the primary data recording, either by the researcher for the purposes of reliability analyses or by other researchers to verify the results. Furthermore, manual data entry from check sheets is time consuming and error prone. Pipeline A represents the classical approach in ethology and field biology.

Pipeline B makes use of computer-based technology for data capture and coding. The data is captured with digital video recording. The video is then observed and manually coded by the researcher using event-recording software. One advantage of pipeline B over pipeline A is that the data is automatically recorded in a rich form that permits the measurement of intra- and inter-observer reliability and independent verification by other researchers. A second advantage is that the time-consuming and error-prone process of manual data entry is eliminated by coding the data directly onto a computer. A third advantage is that the raw video data is available for future studies using different metrics, or for teaching purposes. A disadvantage of pipeline B is that manual coding is time consuming and needs to be checked for intra- and inter-observer reliability. Pipeline B represents a more modern ethological approach.

Pipeline C involves the collection of tri-axial accelerometry data from loggers attached to the birds. For a subset of the captured data, synchronised video is additionally collected. The video is then coded manually to identify roosting, using the same behavioural definition adopted in pipelines A and B, and the results are used to manually label time points at which roosting is occurring in the acceleration data file. A subset of the labelled dataset is used to train a machine-learning (ML) algorithm to classify acceleration data as either 'roosting' or 'not roosting'. The resulting algorithm is validated with the unused subset of the labelled data. The algorithm is then used to classify the unlabelled accelerometer data. An advantage of pipeline C is that by automating the extraction of the roosting metric from the acceleration data, every second of the data collected can be coded. Another advantage of an automated approach of this type is that the same algorithm is used throughout, improving the reliability of coding. The entire pipeline is open to independent verification if the raw data files and ML algorithms are published. Pipeline C represents a new generation of approaches to big data in behavioural science, sometimes referred to as **behavioural informatics**.

three pipelines use technologies that are described later in the chapter, each of which has advantages and disadvantages.

A universal feature of data pipelines is that the dimensionality (or complexity) of the data is progressively reduced. For instance, video-only recording loses information about vocalisations, three-dimensional movement and physiological variables. Accelerometry loses information about everything except body movements. Manually coding one behaviour with time sampling loses information about all other behaviours and only retains information about the target behaviour at the chosen sample points. Dimensionality reduction is an inevitable consequence of moving from the richness of raw behaviour to one or more specific metrics.

In the rest of this chapter, we will follow the data pipeline step by step as far as the production of metrics. The final stage, in which metrics are used to test hypotheses, is covered in Chapter 11 on data analysis.

8.3 Data Capture

Many different technologies are available for capturing behavioural data. The choice of technology will depend largely on the hypothesis under test. Most behaviour is multimodal, involving body movements, vocalisations (possibly including infrasound or ultrasound), written language in the case of humans, and potentially other modes of communication such as changes in body colour, chemical signals and even, in some species, electrical signals. The right technology must be chosen to capture the relevant modalities.

All data capture technologies involve some loss of information, and the information that is *not* captured needs to be considered. For example, sound recording technology optimised for human hearing may be unable to capture the ultrasonic vocalisations of rats. Similarly, video optimised for human vision will not capture the UV plumage 'colours' of birds.

More than one technology may be needed; for example, time-synched video may be required for the initial interpretation and coding of accelerometer data (as in pipeline C). The use of multisensory data loggers (**biologging**) is increasingly common because it provides more information than one source alone. For instance, the simultaneous measurement of satellite-based geolocation in conjunction with atmospheric pressure makes it possible to locate the feeding sites of diving birds in the open ocean.

Data capture technologies can broadly be categorised along three distinct dimensions: (1) the richness of the data captured; (2) the extent to which data capture is automated; and (3) the extent to which data is coded into metrics at the time of data capture. There is generally a trade-off between the richness of the data captured and the need for subsequent coding: rich data types such as video require extensive processing to code behavioural metrics, whereas simple data types such as heart rate do not. In some cases, the researcher can choose between capturing data in a rich uncoded form (as in pipeline B) or coding it at the time of capture (as in pipeline A).

Coding data at the time of capture has the advantage of forcing the researcher to specify the behavioural metrics before starting data collection. It is also efficient because no subsequent coding is required. However, coding at the time of capture means that the resulting dataset cannot be used later to answer different questions that require different aspects of behaviour to be coded. A further downside is that the reliability of coding cannot subsequently be checked independently.

The behaviour in many standardised behavioural tasks is captured in a form that requires little or no subsequent coding because the required metrics are elicited from the subject and recorded automatically. For example, apparatus such as the radial-arm maze for measuring memory in rodents often includes dedicated technology to extract metrics such as the number and timing of arms visited.

When observing spontaneous behaviour, coding is more often done manually. In this case, the researcher must decide whether to code data at the time of primary data capture or later. Given the very low cost of digital storage, it makes sense to record data in a relatively rich uncoded form where possible. The uncoded data can then be made available for independent verification and other purposes. This protects the researcher against suspicions of random error, bias or deliberate falsification of data, and allows the data to be recoded and reanalysed differently at a later date.

Given a choice, automated data capture is usually preferable to manual recording. Automation reduces the time taken to capture data and is therefore likely to increase the amount of data it is possible to collect. It also eliminates problems inherent in manual recording, particularly observer drift and poor inter-observer reliability. However, automating data capture may require expensive equipment or customised software, making it unachievable in low-budget studies.

The automation of coding is less straightforward. Despite advances in technology, some coding tasks are still done faster or more reliably by humans than machines. Moreover, good-quality manual coding of behaviour is often necessary to train machines to perform automated coding (as in pipeline C).

In the following sections, we outline the main technologies for capturing and coding behavioural data. Table 8.1 compares examples of these technologies in terms of data richness, automation of data capture and the extent to which data is coded at the time of capture.

8.3.1 Video

Video is a good way of capturing rich data on body movements and the interaction of subjects with their physical and social environments. A major advantage is that a single camera can cover a large area and multiple subjects. Video may be the only option if the behaviour is very fast or complex, or involves many individuals. It is also the primary means of data capture in many dedicated measurement tools, such as eye-movement tracking and automatic vehicle recognition.

Although video captures rich data, some information is lost. In particular, three-dimensional movement is hard to reconstruct accurately from a two-dimensional video. Three-dimensional cameras, which additionally capture information about depth, can solve this problem. Another limitation of video is the restricted field of view, which can result in contextual information such as group size being lost.

Specialised video cameras are available for particular applications. Monochrome cameras typically have greater light sensitivity than colour cameras and tend to be better for low light conditions. Infrared cameras can record warm-blooded subjects in complete darkness. High-speed cameras can be used for the analysis of very fast behaviour, UV-sensitive cameras can record visual signals in the UV part of the spectrum, and thermal imaging cameras can identify physiological indicators of stress and emotional changes. Programmable motion-sensitive cameras capable of recording still images or video (**camera traps**) are used in the laboratory and field.

Table 8.1 Comparison of various technologies for data capture

Technology	Richness	Automation	Coding
Three-dimensional colour video	Very high	High	None
Audio recording of an interview	High	High	None
Two-dimensional monochrome video	High	High	None
Text transcript of interview captured with automatic voice recognition	Moderately high	High	None
Text transcript of interview transcribed manually	Moderately high	None	None
Verbal description of behaviour recorded manually by researcher	Moderate	None	Low/ none
Tri-axial accelerometry	Moderate	High	None
Satellite-based geolocation (e.g. GPS)	Moderate/ low	High	None
Check sheet (paper and pencil)	Moderate/ low	Low	High
Psychological rating scale (online)	Moderate/ low	High	High
Live computer-based event recording	Moderate/ low	Moderate	High
Computer-based reaction time task	Low	High	High
Activity monitor in cage (e.g. running wheel)	Low	High	High
Heart rate	Very low	High	High
Smartphone poll with yes/no question	Very low	High	High

8.3.2 Sound

High-quality sound recordings are required when animal vocalisations or human spoken language are the focus of a study. Obtaining good recordings in the field can be challenging. In order to maximise the signal-to-noise ratio, the microphone should be placed as close as possible to the vocalising subject and kept away from other noise sources and acoustically reflective surfaces. A directional microphone helps to attenuate noise and echoes. Specialised recording and playback equipment is needed when working with ultrasound (e.g. in bats or rats). Programmable sound-sensitive audio recorders are available for use in the laboratory or field.

8.3.3 Text

Text of spoken or written language is an important primary data source in human behavioural research. Text is captured in various ways. Pre-existing text in digital form can be obtained from publications, emails, internet chat rooms and social media posts. Responses to specific questions or prompts can be elicited from participants in focus groups, interviews, structured interviews or questionnaires. If the primary data is spoken or recorded on paper, it will need to be transcribed into digital form before it is analysed. Good software exists for automatic voice recognition and for machine reading of digital images of printed text and handwriting (via **optical character recognition**). Manual transcription is time consuming and potentially error prone, and should be avoided where possible, although it may be necessary in difficult cases such as historical records in archaic handwriting.

8.3.4 Rating Scales

Specific information can be elicited from human subjects using specialised questionnaires designed to yield **rating scales** (also known as **Likert scales**). Rating scales are commonly used in psychology and the social sciences to assess constructs such as depression, subjective well-being, pain, personality and mood. A rating scale comprises a series of **items**. For each item, the subject is typically required to indicate their level of agreement or disagreement with a statement. The subject answers by choosing one of a series of ordered but discrete statements (e.g. strongly agree, agree, neutral, disagree, strongly disagree) or with a **visual analogue scale** – a horizontal line indicating a continuum from strongly disagree to strongly agree, or from least to most. The responses to items are generally combined to produce an overall score according to a defined method.

8.3.5 Other Modalities

Some communication occurs in sensory modalities that require specialist instruments. For example, the best way to measure body colour is with a spectrophotometer or hyperspectral imaging camera. Measuring the electrical signals emitted by a weakly electric fish requires an electronic amplifier. Similarly, analytical instruments are needed to study chemical signalling by identifying chemical compounds deposited in the environment (e.g. scent marks or odour trails) or released into the air (e.g. airborne pheromones). The technical difficulties in measuring aerial chemicals have led to this modality of behaviour being largely ignored in favour of vision and sound. However, technologies such as the electroantennogram and proton transfer reaction time-of-flight mass spectrometer are able to measure aerial chemicals in real time.

8.3.6 Measuring Physiological Correlates of Behaviour

A supplement or alternative to measuring behaviour directly is to measure physiological correlates of the behaviour or the cognitive and emotional processes underlying it. Physiological variables such as breathing rate, heart rate, blood pressure and skin conductivity can be measured in the laboratory with equipment similar to the polygraph used in lie-detection tests. Newer technologies such as facial thermal imaging, laser Doppler vibrometry and voice stress analysis are able to measure similar physiological variables at a distance, without having to attach wires to the subject. Using physiological measurements to make inferences about a subject's behaviour or mental state is fraught with methodological issues that we do not have the space to explore here.

8.3.7 Autonomous Sensors Attached to Subjects

Another way of capturing behavioural data is to attach autonomous sensors that measure specific variables related to the subject's behaviour. The development of miniaturised autonomous sensors has greatly expanded the range of data that can be collected, particularly from subjects in their natural environment or where other forms of recording are impractical. Versions of such technology exist both for humans (e.g. smart watches and smartphone apps) and animals. Wearable sensors have the big advantage of not requiring the subjects to be visible to an observer or to remain within a limited space.

Sensors can measure a range of behaviourally relevant physiological and environmental variables, including heart rate, body temperature, blood pressure, air temperature, light level and depth in water. Accelerometers that measure changes in body velocity can capture body movements and postures. Tri-axial accelerometers measure acceleration in the three orthogonal dimensions, corresponding to the surge (forwards and backwards), heave (up and down) and sway (side to side) movements of the body, with some devices capable of recording data many times per second. Geolocation is possible via telemetry, geomagnetic loggers and satellite-based global positioning systems (GPS).

Sensors for many of these variables are small enough and light enough to be deployed in tags on small birds and mammals. Tags containing sensors can be attached externally with collars, bracelets, leg rings or harnesses, or they can be implanted inside the subject. Data may be recorded on the device itself or transmitted to a receiver, periodically or in real time. Tags that transmit data typically require larger batteries, making them heavier. A major consideration is the ease with which tags can be recovered to retrieve stored data; if recovery is likely to be problematic, it may be better to use a transmitting tag.

With humans, the smartphone is effectively a multifunctional sensor attached to the subject. Smartphones can capture data on a range of behavioural types, including activity, sleep quality, long-range movements and inter-personal interactions. Their utility in capturing behavioural data from very large samples of freely behaving people came to the fore during the COVID-19 pandemic.

8.3.8 Fixed Sensors in the Environment

Installing fixed sensors in the environment that detect the presence of a subject is a relatively low-tech way of recording some behavioural metrics. For example, the activity of an animal in a cage can be recorded using running wheels, arrays of light beams transecting the floor space, or micro-switches on food hoppers or under perches. Similarly, human movements along a road or path can be recorded by fixed sensors that count and classify vehicles, bicycles and pedestrians. Speed and direction of travel can be recorded by installing two closely spaced sensors along a path. Other types of behaviour can be measured by placing sensors in relevant locations – for example, by recording when an animal accesses a food hopper.

Some studies require individual animals to be identified (see section 9.2). This can be achieved by attaching unique electronic tags that are detected by readers at food hoppers, nest entrances or other locations of interest. **Radio-frequency identification (RFID)** tags contain a miniature radio receiver and transmitter; when triggered by an interrogation pulse from a nearby RFID reader, the tag transmits its identifying number back to the reader. RFID tags are of two basic types: **passive tags**, also known as PIT (passive integrated transponder) tags, which are powered by energy from the RFID reader's radio emissions; and **active tags**, which are powered by an internal battery, enabling them to be read at a greater distance from the reader. Passive RFID tags are generally smaller than active tags and can remain operational for many years. RFID tags can be attached to a collar or leg ring, while the smaller tags can be injected subcutaneously.

Another type of tagging technology uses printed **barcode**-type images, which are fixed to subjects and read by high-resolution cameras. The advantage of barcodes over RFID tags is that they are cheap and can be very small (<1 mm^2), making it feasible to collect data from hundreds of tiny animals. They can be used with almost any species that can be kept in a relatively flat and well-lit arena. This technology has been used to investigate, for example, the social structures of ant colonies, by inferring social interactions from the spatial and temporal proximity of individuals [84].

8.3.9 Computerised Behavioural Tasks

Many behavioural studies use computerised tasks in which the human or non-human subjects interact directly with bespoke software by means of touchscreens, keyboards, joysticks, levers, nose-poke holes, pecking keys or pedals. Such tasks are commonly used to measure aspects of learning, memory and decision making. Computerised tasks require human participants to be instructed and animal subjects to be trained, which may be time consuming. Once the subjects are trained, however, computerised tasks allow huge amounts of data to be collected automatically.

Computers are used both to run experiments (e.g. by presenting stimuli and administering rewards or punishments) and to record data (e.g. by measuring reaction times, recording choices or counting lever presses). The software can be custom written in standard programming languages such as MATLAB or Python, or derived from specialised software packages called **experiment generators** [85]. These packages contain libraries of standard tasks that are easily modified [86, 87]. Specialised software and hardware are available for running learning experiments with animals. For human studies, smartphones offer a cheap and effective means of polling participants about their current feelings or behaviour in real time. Numerous apps are available for smartphone polling.

8.3.10 Event Recorders

An event recorder is a flexible computer-based tool for manually coding behaviour. Numerous software packages are available for converting standard laptops, tablets and smartphones into event recorders, some of which are freeware [88, 89]. Event recorders are commonly used to code video data and live behaviour in the field.

Event recorders are capable of recording behavioural states and events using continuous recording or time sampling (see Chapter 7). Observations are recorded as key presses on a standard keyboard or other input device. Typically, different keys are assigned to particular categories of behaviour or particular subjects. A standard keyboard can easily be customised with labels to facilitate complex behavioural coding.

8.3.11 Check Sheets

A check sheet is a simple, cheap and flexible tool for manually recording observations. With a properly designed check sheet and some practice, a lot of information can be recorded reliably and with reasonable accuracy. Check sheets can record both standardised behavioural tests and spontaneous behaviour, using any of the recording methods described in Chapter 7. Many of the functions of a computer-based event recorder can also be implemented with a check sheet.

The basic design of a check sheet is a grid, with columns denoting different categories of behaviour or types of data, and rows denoting successive trials or sample intervals. At the end of each trial or sample interval, the observer starts recording on the next row of the check sheet. A simplified example of a check sheet for coding spontaneous behaviour is shown in Figure 8.2.

The number of columns will depend on the number of categories recorded and can be reduced by using different symbols within each column to represent different categories or subcategories of behaviour. For example, within a column labelled 'vocalisations', different symbols can signify particular types of vocalisation. Symbols can also identify individual subjects, such as the initiator and recipient of a social interaction, as shown in Figure 8.2. It may help to leave one or two blank columns for new categories

Mother: Gladys Infant: Mabel Group: B	Date: 8 Aug 20 Start time: 14:30 Observer: PHM Temp: 19°C		Observation session no. 3		Page 1	
Ventro-ventral contact	Groom	Approach	Leave	Eat	Remarks	
✓						
✓						
✓			M			
		I				
	M					
	M				Grooming belly	
	M					
			I	I		
				I		

Figure 8.2 A simplified check sheet, in this case for recording the behaviour of a rhesus monkey mother and her infant. Five categories of behaviour were recorded, using continuous recording (CR), instantaneous sampling (IS) or one–zero sampling (1/0): 'Mother and infant in ventro-ventral contact' (IS); 'Groom' (IS); 'Approach' (CR); 'Leave' (CR) and 'Eat' (1/0). In this example, the mother and infant were initially in ventro-ventral contact. The mother (M) then left the infant. The infant (I) then approached the mother and the mother groomed the infant. Finally, the infant left the mother and started to eat. Only the top part of the check sheet is shown.

that are added during preliminary observations, or to record additional information that may later be useful in interpreting the results.

For studies that make use of focal sampling, scan sampling and ad libitum sampling, different parts of the same check sheet can be set aside for categories recorded using the three sampling rules. Similarly, all three recording rules (continuous recording, one–zero sampling and instantaneous sampling) can be implemented on the same check sheet. With instantaneous sampling, the behaviour is recorded by marking 'on the line' at each sample point. With one–zero sampling, a mark is made 'in the box' when the behaviour first occurs. For continuous recording of frequencies, each occurrence is marked within the corresponding sample interval. The exact frequencies of events, as well as their sequence, can be recorded relatively easily on a check sheet. With one–zero sampling and instantaneous sampling, a beeper (a device that produces a regular time signal) is needed to denote the sample intervals.

Recording precise durations is harder with a check sheet than with an event recorder because the times at which the behaviour started and stopped must be captured. A cruder way of recording approximate durations is to mark within the relevant sample interval (the current row) whenever the behaviour stops and starts. This is easier than reading a clock and noting the actual times, but timing is only accurate to the nearest sample interval. If the sample interval is short relative to the average bout length of the behaviour, then this method may be adequate.

Data from paper check sheets must eventually be entered into a computer, which can be time consuming and error prone. Where possible, it makes sense to record data directly onto an electronic check sheet on a laptop or tablet in order to avoid manual data entry.

8.3.12 Verbal Descriptions of Behaviour

Researchers sometimes record verbal descriptions of behaviour, in the form of written notes or by dictating into an audio recorder (e.g. a smartphone). Verbal descriptions can be a useful flexible supplement to more structured recording methods, especially during informal pilot observations and for recording rare events. In some cases, dictating a verbal description is a practical way of recording complex behaviour involving multiple categories that cannot reliably be coded directly onto a check sheet or keyboard.

8.4 Data Processing

If behavioural metrics are not extracted at the time of data capture, then some subsequent data processing will be required to code the data. Rich primary data sources such as video, sound and text almost always require coding. Coding can be manual or automated in some way.

Manual coding of rich behavioural data involves the researcher inspecting the data and using their judgement to identify behaviour and assign it to categories by applying predefined rules (see Chapter 6).

Automated coding of rich behavioural data is generally performed by some type of AI – digital technologies that enable machines to perform complex human-like tasks such as pattern recognition and decision making. One advantage of automated coding over manual coding is that the coding rules remain consistent over time. Automated coding is also much faster than manual coding, making it a practical necessity when the volume of data is very large or when results are required quickly. The growing capacity to acquire big data in behavioural science has led to rapid developments in AI technologies for automated coding.

AI-assisted methods for automated coding of behavioural data can be divided into two broad types, which we refer to as **rules-based** approaches and **machine-learning (ML)** approaches. In rules-based approaches, a human expert pre-defines the rules for coding behaviour, which are then implemented by a computer (Box 8.2), whereas in ML, the computer learns the behavioural coding rules for itself by discovering patterns within the data supplied to it (Box 8.3). ML has been described as a form of automated statistical analysis. It has been credited with solving difficult problems, such as automated speech recognition, that had previously eluded rules-based approaches. A new generation of so-called **deep learning** algorithms, based on complex multilayered artificial neural networks, are becoming common in behavioural science.

Even when most of a dataset is coded automatically, some manual coding may still be required in order to validate and further develop the automated method. Thus, many data pipelines include both manual and automated coding (as in Box 8.1 pipeline C).

Researchers can choose from a variety of software packages to assist in the coding of behavioural metrics from rich primary data. Freeware that performs essentially the same functions as expensive proprietary packages is

Box 8.2 Rules-based approaches to coding behaviour

Rules-based approaches depend on human subject matter experts to explicitly define algorithms for coding the behaviour. The expert researcher decides what data types are likely to be most informative for identifying a given behaviour and what thresholds to set for variables indicating the presence of the behaviour.

To take a very simple example, if the task is to code the frequency and duration of vocalisations produced by a subject, the researcher could define a vocalisation as occurring whenever the amplitude of the waveform obtained from a sound recording exceeds a threshold chosen by the experimenter. The value of this threshold is decided by the researcher based on some intrinsic biological or psychological criterion, or by manual examination of the variation in a subset of the data. Once the rules are set, a computer algorithm embodying these rules steps through the entire dataset, identifying the start and stop of all vocalisations. The algorithm could be modified to identify specific types of vocalisations by adding additional rules – for example, that the vocalisations must lie within defined ranges of duration and frequency. More complex vocalisations could be identified by specifying, for example, how the frequency changes over time. Similar approaches are widely used in the analysis of movement data derived from attached sensors. For example, a rules-based approach has been used to detect when a penguin is walking, based on characteristic periodic oscillations in the rate of change of heave acceleration [90]. Rules-based approaches can incorporate tolerance ranges to accommodate variations in the timing and amplitude of instances of a behavioural category.

A related approach for detecting more complex patterns in data relies on **template matching** [91]. The researcher supplies the algorithm with a template in the form of an archetypal example of data corresponding to the target behaviour, such as a segment of a spectrograph or an accelerometer trace corresponding to a stereotyped vocalisation or movement. The algorithm slides the template across the dataset, computing the similarity between the template and the data at each position. A match is recorded whenever this similarity exceeds a threshold decided by the researcher. Template matching works best for behaviours that have a stereotyped temporal pattern.

An advantage of rules-based approaches is that, where reliable rules can be found, the coding process is transparent and easy to communicate. However, there are many behavioural coding problems for which it is difficult to identify reliable rules.

Box 8.3 ML approaches to coding behaviour

ML is a form of AI in which an algorithm improves its performance at a task through exposure to training data, without being explicitly programmed to perform the task. The algorithm learns, in the sense that its outputs are modified by its experience. When coding behavioural data, a typical task would be to identify whether a specific behaviour is present in a dataset. The training would involve presenting the algorithm with multiple examples of the data. A successful ML algorithm will generalise to novel examples of the same type of data and make accurate classifications of previously unseen data without the researcher specifying how this is done.

ML is a powerful tool for automating the extraction of simple metrics from complex behavioural data. It is increasingly used in behavioural science to code big data derived from spoken language, text, video and sensors such as accelerometers and GPS tags [92]. The development of accessible software packages has facilitated the uptake of these techniques by researchers who are not themselves experts in computer science [93, 94].

There are two basic types of ML. In **unsupervised ML**, the training phase involves giving the algorithm examples of data but no information on the desired output. Unsupervised ML methods are used to discover statistical patterns in unlabelled data, such as the grouping or clustering of data points. In **supervised ML**, the training phase involves giving the algorithm examples of data along with instructions about the desired output. For example, the images or segments of accelerometer data that are used for training would be labelled manually by the researcher according to whether they contain the behaviour of interest. A drawback of supervised ML is the requirement for a labelled training dataset (sometimes called a calibration dataset), which can be time consuming to produce and may require expert input.

The effectiveness of any ML algorithm is limited by the quality of the dataset used to train and test it. The training data should reflect the breadth of the full dataset to be analysed and should contain examples of the full range of the behaviour to be coded – in other words, different views of the same behaviour, different individuals performing the behaviour, different luminance conditions for video recordings, and so on. Published papers based on ML should include full descriptions of how the training data was collected and, if relevant, coded.

When ML is used to code behaviour, the coding is repeatable because the same algorithm can be used by other researchers. However, it may be impossible to specify how the algorithm produced its outputs. An ML algorithm is often a 'black box' that cannot be reverse engineered – an issue sometimes referred to as the problem of explainability. This lack of transparency about the inner workings of ML can lead to failures of generalisation if the training data is biased. In an often-repeated story about an early military application of ML, a neural network was trained to code images as either containing tanks or no tanks. The algorithm generalised very well to novel images from the original dataset. But when it was used on a new dataset, the algorithm performed no better than random guessing. It transpired that all the original 'tanks' photos were taken on a sunny day, whereas the 'no tanks' photos were taken on a cloudy day. The network had learned something about the illumination of the images, as opposed to the features of tanks.

available. Some studies require customised software to perform certain types of non-standard coding of data, in which case researchers can use tools that facilitate the customisation. Some freeware has open-source code for those who wish to modify it, and libraries of algorithms are available in generic programming languages such as MATLAB, Python and R.

In the following sections, we introduce some of the technologies available for assisting researchers in coding the main types of behavioural data.

8.4.1 Coding Video Data Manually

Behavioural metrics are commonly extracted from primary video data using event-recording software that enables the researcher to manually code the behaviour directly onto a computer. Most event recorders allow integrated video playback, making it easy to control the position in the video, the speed of playback and the sample interval (if relevant). This allows the coder to review sections of video to check for coding errors, or repeatedly view ambiguous sections before coding. Some event recorders allow time-synched video or audio files to be displayed and coded simultaneously, which can be helpful if more than one view of the same behaviour, or more than one modality, is needed to code the behaviour reliably.

Another common application is the manual annotation of objects of interest within a video file – for example, instances of an animal performing a particular behaviour. Specialised software allows the researcher to mark and label areas of interest on individual video frames. Labelled video is required as training data for supervised ML.

8.4.2 Coding Video Data Automatically

Machine vision is the use of AI to automate the extraction of information from video data. Technologies based on machine vision can improve the speed and reliability with which behaviour is coded from video [95].

It is not always necessary to mark or identify individual subjects in order to extract useful behavioural metrics from video data. If the subjects contrast sufficiently with their background (e.g. pale chickens on darker litter), an **occupation index** can be obtained by measuring the proportion of pixels above or below a defined threshold, either at the level of the entire image or for focal regions within an image. Another attribute, known as **optical flow**, is based on detecting the rate of change of brightness in the pixels over time. The statistical properties of optical flow are affected by the behaviour of the subjects and can therefore be used as a behavioural metric. For example, aspects of the optical flow generated by the collective movements of flocks of broiler chickens are correlated with changes in the lameness of individual birds, providing an automated method for detecting signs of poor welfare in commercial flocks [96].

More sophisticated machine vision technologies are capable of identifying a focal subject or subjects within an image or video frame. The process of differentiating a subject from its background, known as **segmentation**, is relatively easy for a single, highly contrasting subject, such as a black mouse in a white arena, but it becomes harder with multiple similar subjects or subjects that are well camouflaged. The location of a subject with respect to fixed features of its environment, such as food and water sources, nests or perches, can be used as a proxy for corresponding behaviours such as eating, drinking and resting.

Once a subject has been identified, its position in subsequent video frames can be **tracked** automatically over time. A range of useful metrics can be obtained from tracking data, including speed and acceleration of movement, distance travelled and the straightness of a path. Such metrics can in turn form the basis for detecting specific behaviours such as walking, running, play or aggression.

Feature extraction involves identifying particular parts of the subject's body, such as the head and rear end of an ant, or the eyes, nose and mouth of a human face. The identification of body parts enables the automated recognition of more categories of behaviour. For example, two ants that come into contact may be coded as engaging in a social interaction if, and only if, it is their heads that touch [84].

Pose estimation uses the spatial relationships between an individual's body parts to infer different postures or expressions. For instance, the relative positions of the major joints of a monkey vary systematically according to whether it is sitting, reaching or walking. Similarly, the relative positions of the features of a human face vary according to whether the subject is smiling or scowling. The ways in which the spatial relationships between features change over time can in turn be used to identify more complex behaviours comprising sequences of poses or expressions. The gold standard for pose estimation involves fixing reflective markers to key joints or other locations of interest on the subject's body. However, advances in ML-based feature extraction have made it possible to achieve **markerless pose estimation** in a range of species [94–97].

8.4.3 Coding Sound Data

The most salient features of a sound recording can be quantified in terms of changes over time in amplitude (loudness) and frequency (pitch). Sound recordings can be visualised in various ways: a simple audio **waveform** plots amplitude against time, while a **spectrogram** (also known as a sonograph, voiceprint or voicegram) plots frequency against time and represents amplitude by the darkness of pixels. Waveforms are useful for conveying basic information, such as whether someone is speaking or not, or the overall energy content of a bird's song. Spectrograms contain more detail about the type of sound, such as the presence of particular motifs in birdsong. An experienced researcher can read a sound spectrogram like a musical score. Software packages for analysing sound usually include tools for facilitating manual and automated coding of sound data.

Machine hearing is the sound equivalent of machine vision. The best-known implementation is the software for **automated speech recognition (ASR)** on smartphones and digital assistants such as Siri and Alexa. In ASR, the spectrogram of speech is broken down into **phonemes**, the unique sounds that make up human language. ASR software uses statistical probability analysis to deduce words from sequences of phonemes and, from there, complete sentences. ML has been central to the development of accurate algorithms for ASR. One important application of ASR is converting spoken language into digital text.

8.4.4 Coding Text

Text analytics is the process of converting unstructured text into meaningful data for qualitative and quantitative analysis. Many software packages have been developed for coding digital text, which allow text to be imported directly from social media, emails, web survey platforms and other online sources. Packages exist to facilitate manual and automated coding. Manual coding involves marking, annotating and classifying meaningful passages identified by the researcher. Once passages have been identified, other tools are capable of distinguishing recurrent themes and visualising relationships between passages. Automated search functions using **Boolean queries** (search terms with logical operators such as 'AND', 'OR' and 'NOT') can locate and count key words or phrases for quantitative content analysis. ML is increasingly used on text to identify complex categories that are not amenable to rules-based approaches, such as identifying communications about terrorism or criminal activity.

8.4.5 Coding Movement and Location Data

Movement-based data can be processed and coded in various ways. Data from accelerometers attached to subjects can be visualised by plotting acceleration against time, with tri-axial accelerometers producing three graphs representing surge, heave and sway. Such data can be used to compute metrics related to overall activity and energy expenditure [98]. Some commercial devices automatically produce metrics of this type. Accelerometer data can also be used to identify specific behaviour patterns such as walking or soaring [99]. Accelerometer data is amenable to both rules-based and ML techniques for identifying and quantifying behaviour, and software packages are available to facilitate this [90–93].

Movement data captured from GPS tags can be visualised by plotting spatial location (latitude and longitude) against time. The tracks yield a range of behavioural metrics, including speed and direction of movement and the tortuosity of the path. ML can also be used to identify particular types of behaviour, such as the diving of seabirds [100]. The automated coding of acceleration or GPS data often requires some concurrent data from another reliable source in order to develop the rules for classification or for training an ML algorithm (see Box 8.1, pipeline C).

8.5 Summary

- High-quality behavioural data can be recorded using cheap and simple technologies such as check sheets and sound recorders.
- Advances in technologies for data recording have made big data available to behavioural scientists, which in turn has stimulated the development of AI technologies for automated data processing.
- A data pipeline describes the workflow of data recording, processing and analysis, including details of the technologies used in each step.
- The choice of technology for capturing behavioural data will depend on the research question and the resources available, the quantity of data required, where the data is to be collected, the amount of interaction with subjects and the likely impact of the technology on the subjects and their environment.
- Data that is initially recorded in a relatively rich form will require subsequent processing to code behavioural metrics.
- Coding of data can be either manual or automated using rules-based approaches and machine learning.

9
Individuals and Groups

Animals interact with other members of their species in interesting and important ways. Some animals live in stable pairs or larger groups, whereas other groupings are more dynamic, varying in size or composition with the location, season or type of activity. The size of a group in which an individual lives and the nature of their social interactions have profound effects on their biology and psychology. For these and other reasons, the causes and consequences of group living and social interactions are major areas of research.

Many aspects of social behaviour are studied by measuring the behaviour of focal subjects sampled from a population. Subjects may be observed interacting in their natural environments, or measured with standardised behavioural tests or questionnaires, as outlined in previous chapters. Much of what scientists know about communication, cooperation, aggression, mate choice and parental care has been learned in this way. However, some research questions require the study of whole groups of individuals interacting in their natural environments. The collection of data on the social relationships of whole populations of individuals interacting naturally, known as **reality mining**, has been greatly facilitated by advances in technology for identifying individuals and quantifying their interactions [101].

The interactions between individuals within a group can have consequences at the group level. The field of **collective behaviour** looks at how entire groups move through space and how they respond to perturbations such as encountering a physical barrier, the arrival of a predator or the introduction of new individuals bearing valuable information [102]. Research topics include such things as the waves of stop–start movements observed in columns of heavy traffic, the dynamic patterns formed by flocks of starlings or shoaling fish, and the emergence of leadership in human and animal groups. Interacting individuals generate **emergent properties** at the level of the group that are not evident from studying individual behaviour. These group-level phenomena are sometimes the most informative unit of measurement.

In the following sections, we outline the main methods for studying social interactions of individuals within groups and the consequences of these interactions at the group level.

9.1 Groups

We define a group in general terms as an aggregation of individuals. Examples include a troop of baboons foraging on the savannah, a flock of starlings roosting in a tree, a table of students chatting in a canteen and a queue of people waiting at a bus stop. Being observed in a group on a single occasion does not necessarily imply the existence of biologically meaningful relationships or interactions between group members. However, if the same individuals are repeatedly observed together, this information could imply the existence of more stable and potentially meaningful groups [103]. The **gambit of the group** is the assumption that if two individuals are repeatedly observed in the same group, then they have some sort of social affiliation and interact more with each other than they do with others outside the group.

9.1.1 Identifying Groups

What constitutes a group? Where does one stop and another start? In practice, groups are often defined by assessing how individuals are distributed in space and observing the relative distances between them. Those within a certain distance are defined as being in a group. The rules for defining a group should be as explicit as possible, using clear criteria for membership. For instance, an individual might be defined as part of a group if their nearest neighbour, defined as the one whose head is closest to the head of the target individual, is within, say, two body lengths. The appropriate distance will, of course, depend on the species, the circumstances and the research question.

The members of a group cannot always be directly observed simultaneously. Group membership must sometimes be inferred from indirect data – for example, the temporal sequence with which individuals visit a fixed location such as a feeder. A simple approach is to regard individuals as belonging to the same group if they are recorded at a given location within a fixed time window. A problem with this approach is that the time window chosen is likely to influence the result. Significant associations will be missed if the window is too small, whereas meaningless associations will be inferred if it is too large. To get around this problem, statistical clustering algorithms can be used to find the best-fitting temporal boundaries of groups [104]. This method is, of course, based on an assumption that true groups do exist.

With non-human species, groups are usually determined by physical proximity. But this is not always the case for humans, for whom many forms of communication, such as email, phone and social media, are independent of the physical distance between interacting individuals. Accordingly, some human groups are defined by the frequency or duration of communication rather than physical proximity. Social network analysis provides methods for defining groups according to social relationships (see section 9.3).

9.1.2 Estimating Group Size

Group size is an important metric with implications for individual behaviour [105]. It may be possible to count the individuals in a group directly, provided the group is not too large. Counting can be automated in situations where individuals contrast sufficiently with their background (which is relatively easy to arrange in the laboratory) or if individuals are marked in a way that can be detected automatically (e.g. with camera-readable barcodes).

When groups are very large, counting may not be feasible, and group size must instead be estimated. It may be sufficient to classify groups into size categories such as 1, 2, 3–5, 6–9, and 10 or more. A trained observer should be able to identify such categories quickly and reliably. If more accurate counts are required, a common approach is to overlay an aerial photograph with a grid and count the number of individuals in a representative sample of grid squares. The group size can then be estimated by multiplying the average number of individuals in a square by the total number of squares occupied by the group. For humans, good estimates exist for the average number of people per square metre in crowds of different overall densities.

The number of people in a loose crowd is typically around 1 person m^{-2}, in a solid crowd is around 2 people m^{-2} and in a very dense crowd is around 4 people m^{-2}. Thus, to estimate the number of people in a crowd, it is only necessary to estimate the overall density (loose, solid or very dense) and the area occupied [106]. For groups that are moving, such as mass migrations or demonstration marches, it may be feasible to estimate group size by counting the number of individuals passing a given point per unit of time. Estimates of crowd size may be corroborated with behavioural proxies such as public transport fares, volumes of litter generated and portable toilet use.

As with many other aspects of behaviour, AI is increasingly used to facilitate measurement. For example, deep learning has been used to identify, count and categorise the basic behaviour (e.g. moving, resting, eating) of 48 species of animals in 3.2 million camera-trap photos collected in the Serengeti national park [107]. Before this development, the animals in these photos were manually counted by crowdsourced human volunteers, with several people counting each photo to ensure reliability.

9.1.3 Outsider Versus Insider Perspectives on Group Size

In most species, group size is not normally distributed but has a right-skewed distribution, with many small groups, fewer large groups and very few very large groups. This has important implications for the metrics used to describe group size. From an outsider's perspective, the average group size is described by the mean or median group size. However, these metrics do not correctly describe the social environment of the average individual within the group. When group size is right-skewed, the average individual lives in a group that is larger than the average group size. In the hypothetical example in Table 9.1, the average group size is larger in sample 1, but the average individual lives in a larger group in sample 2. For this reason, the term **crowding** is used to signify the average group size experienced by an individual within the group.

Whether the outsider's or the insider's perspective (i.e. average group size or crowding) is correct will depend on the research question. Crowding is clearly not an independent measurement for individuals in the same group, which has implications for statistical analysis (see section 11.7). Free software is available to calculate various metrics of group size [108].

Table 9.1 Example of average group size versus crowding (modified from [108])

	Sample 1	Sample 2
Group sizes	4, 5, 6	1, 4, 7
Mean (median) group size	$\dfrac{4+5+6}{3} = 5\,(5)$	$\dfrac{1+4+7}{3} = 4\,(4)$
Mean (median) group size experienced by an individual (crowding)	$\dfrac{4 \times 4 + 5 \times 5 + 6 \times 6}{4+5+6} = 5.13\,(5)$	$\dfrac{1 \times 1 + 4 \times 4 + 7 \times 7}{1+4+7} = 5.5\,(7)$

9.2 Identifying Individuals

Some research can be conducted without needing to identify individuals, but many studies require individual identification. Focal-individual sampling often requires a reliable identification method, and it is usually important to know whether the same individual is being measured more than once. The analysis of social networks (see section 9.3) and dominance hierarchies (see section 9.4) is impossible unless the majority of individuals in a group are reliably identifiable.

Individuals can be identified in one of two ways: either they are artificially and uniquely marked or tagged, or they are distinguished by naturally occurring variation in their appearance. The identification of individuals can either be done manually or be automated in some way. Automated identification is revolutionising research on social behaviour by allowing the collection of larger datasets.

9.2.1 Marking and Tagging

Many different methods are available for marking animals to make them visually identifiable. These include attaching coloured leg rings, collars, barcodes or other visually readable tags, dyeing fur or feathers, fur clipping, freeze branding, tattooing, toe clipping and ear punching. An alternative to visual identification is to attach, inject or surgically implant miniature radio transmitters or RFID tags.

An advantage of visual or electronic marking is that it enables individuals to be identified more reliably than is generally possible when relying on naturally occurring variation. Moreover, barcodes and RFID tags can be read automatically. However, marking raises both ethical and scientific concerns, and marking methods should be carefully chosen and refined to minimise undesirable side effects. The marking process often involves invasive procedures such as trapping, darting, handling, anaesthesia and minor surgery, all of which have the potential to cause distress or lasting harm. Some of these harms may be mitigated using ingenious methods for marking without capture. For example, territorial hummingbirds have been marked temporarily by remotely spraying their breast feathers with non-toxic coloured ink when they visit a specially adapted feeder.

Even if the immediate harms associated with the marking process are minimised, marks or tags can still have long-term effects. There is mounting evidence that tagged birds have lower survival rates, even when the tags weigh as little as 1 per cent of body weight. It is generally recommended that tags should weigh no more than 3 per cent of body weight in flying animals. Even so, the attachment method, body position and other factors may affect survival. Researchers should always report the details of the tags used, the attachment method and survival data in order to contribute to the growing knowledge base on the safest ways to mark animals [109]. Researchers studying wild animals should also consider how long the marks need to last and whether they will have to be removed at the end of the study or will fall off or fade after a given period.

Artificial marks or tags may have subtle and unanticipated effects on social behaviour. In a famous example, the coloured plastic leg rings used to identify birds were reported to affect mate choice in zebra finches [110]. Although this specific finding is a recent casualty of the replication crisis [111], the general point still stands.

9.2.2 Naturally Occurring Variation

The alternative to artificial marking is to rely on naturally occurring variation between subjects. In some species, individuals have distinctive natural markings. For example, zebras' stripes, like human fingerprints, are unique to each individual. Similarly, gorillas' noses, elephants' ears, the whisker spots of lions, cheetahs' tails and the bills of Bewick's swans, to name only a few, are highly variable. Many animals living in the wild also acquire distinctive marks through injury, such as torn ears, damaged tails, scars or stiff limbs.

Experienced observers can learn to recognise individual animals by naturally occurring features, although this requires patience and practice. Animals that initially appear similar to the inexperienced observer will become perceptually as distinctive as individual humans after a sufficiently long period of observation.

When an observer claims to be able to recognise individual animals, their ability to do so should be validated. One way is to photograph the animals as they are simultaneously identified by the observer. The tester then removes from each photograph any environmental cues that might be helpful in identification and records the identity of each individual. Days or weeks later, the tester presents the observer with the pictures in random order and asks the observer to name each individual [112].

Machine vision is increasingly employed for individual identification. Algorithms can reliably identify individual human faces in digital images, and comparable algorithms have been developed for a number of other species using ML approaches (see Chapter 8). Human face-detection and identification algorithms based on features including the eyes and nose have been adapted successfully for the rhesus macaque and other non-human primate species [113]. Deep learning has been used to develop algorithms for identifying individuals of small bird species, based on their plumage patterns [114]. This approach is well suited to novel individual identification problems like this because it does not require hand-crafted feature extraction (as was done for primate faces); instead, it automatically learns to use the features that are optimal for identifying individuals from training images.

Using naturally occurring variation to identify individuals has the major advantage over artificial marking of being non-invasive. Its disadvantages include the time required for training human observers or developing ML algorithms and the greater difficulty of attaining high levels of reliability. One potential problem with naturally occurring variation is that genetically related individuals may be physically more similar and therefore more likely to be confused, as was found with swans and macaques. Another potential problem is that the features of individuals are likely to change over time. Whether either of these issues is a concern will depend on the research question. Cases of mistaken identity are particularly troublesome for social network analysis.

9.3 Social Network Analysis

Understanding how the individuals within a population interact is fundamental to many areas of behavioural research. Predicting how such things as diseases, fake news, fashions or new ways of obtaining food spread through a population depends on how intimately individuals interact, who they obtain their information from and who they copy.

Social network analysis (SNA) is the framework of analytical approaches used to visualise and describe the network of associations or interactions within a population of identifiable individuals [115, 116]. A social network consists of **nodes** (also known as vertices) connected by **edges** (also known as ties). Nodes usually represent identifiable individuals, although they could represent higher levels of social organisation such as households, villages, companies or species. Edges represent the associations or interactions between each dyad (pair) of nodes. Edges can either be binary, reflecting the presence or absence of a relationship (e.g. whether or not the dyad have had sex or fought each other), or weighted, reflecting the strength of the relationship (e.g. the frequency of interactions). Edges can also be non-directed symmetrical associations (e.g. physical proximity) or directed, when an interaction has an actor and recipient (e.g. grooming, approaching, phoning, boss/employee).

Social networks are visualised in **sociograms**, in which nodes are drawn as points or shapes and edges are drawn as lines or arrows (Figure 9.1). Attributes of nodes (e.g. gender) may be indicated with different shapes or colours. When edges are non-directed, a single line can represent a pairwise relationship between two nodes, whereas when edges are directed, a relationship is represented by one or two arrows. When edges are weighted, lines of different widths are used to represent relationships of different strengths.

SNA is a large and complex topic. In the following sections, we outline the key steps in constructing and analysing social networks.

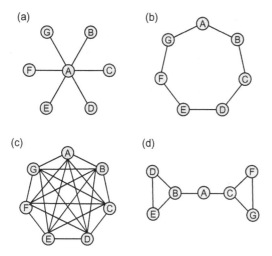

Figure 9.1 Examples of four simple sociograms, each with seven nodes representing seven individuals (A–G) connected by binary, undirected edges. Four different topologies are illustrated: (a) a star network; (b) a circular network; (c) a fully connected network; and (d) a network with two distinct, inter-connected cliques. See section 9.3.2 for more details.

9.3.1 Constructing a Social Network

Constructing a social network requires measuring the associations or inter-actions between each dyad in the population of nodes being studied. The aim is to construct a two-dimensional **sociomatrix** (also known as an association or adjacency matrix; see Figure 9.2b) in which each node appears in both the rows and columns, and each cell contains the data relating to an edge. A sociomatrix is the basic dataset required for all SNA. When edges are non-directed, only half the matrix is required (because the two halves are symmetrical), whereas when edges are directed, the actors are shown in the rows and the recipients in the columns, and the whole matrix is required. The diagonal cells representing the relationship between a node and itself are left blank.

Missing observations in a sociomatrix are likely to affect SNA. Analysis of the impact of missing observations suggests that data collection should focus on maximising the amount of data collected on a subset of nodes, rather than maximising the number of nodes included. As a rough guide, at least 20 observations per node are required. There will often be a trade-off between collecting more observations and making more accurate observations [117].

For many research questions, it is preferable to generate the edges for a social network by measuring a specific type of social behaviour, such as the frequency of grooming or communicating. As individuals will typically vary in their likelihood of being seen, observations should be converted into rates of interaction – that is, the number of interactions observed divided by the duration of opportunity to make such an observation.

Sometimes interactions are hard to observe. An alternative approach is to infer social relationships from more general indices of spatial association, as described below. Of course, spatial association does not necessarily mean that a dyad interacts in a meaningful way, leading to the possibility of false-positive inferences about social relationships. Even so, this approach can still be advantageous if it misses fewer true interactions because even a few missing edges can have a significant effect on network structure.

Associations can be observed directly or inferred from automatically collected data. For example, regular scan samples of a focal individual could record all other individuals in close spatial proximity to the focal individual in a sample interval. Alternatively, associations could be inferred from the temporal sequence of tagged individuals recorded visiting a fixed location (Box 9.1).

How are associations quantified? Any index (metric) of association between two individuals A and B must take account not only of the number of separate occasions that A and B are recorded together but also of the number of separate occasions that A and/or B are recorded apart. In order to calculate such a metric, it is necessary to define what is meant by 'separate occasions' and 'together'. For example, it would clearly be wrong to treat successive scan samples of slow-moving animals taken at 15 s intervals as separate occasions because one observation is effectively the same as the next. 'Together' may be defined in various ways, according to criteria based on proximity in time or space, or simply membership of the same group.

Behavioural observations can be converted into an **association index** that quantifies the likelihood of two individuals being recorded together. The simplest metric of association is the aptly named **simple ratio index**

Box 9.1 Constructing a sociogram from temporal data

Here we illustrate the construction of a sociogram from a temporal stream of individual identifications collected at a fixed spatial location. This highly simplified example shows data collected from a population of RFID-tagged birds visiting a feeder equipped with a tag reader [104]. The raw data points are individual bird identities with the time at which each bird visited the feeder. A short segment of data is shown in Figure 9.2a. A statistical clustering algorithm identified four distinct groups of individuals in this temporal sequence, delineated by the dotted lines. All of the individuals observed in a group are considered as being together. These associations are used to calculate the simple ratio index (SRI) for each dyad. For example, the SRI for the A–B dyad is 2/2 = 1 because A and B were together (in groups 1 and 2) whenever they were observed, whereas the SRI for the A–C dyad is 1/2 = 0.5 because individuals A and C were observed half as often together (in group 2) as they were observed apart (in groups 1 and 3). The SRIs for each dyad are used to populate the

(a)

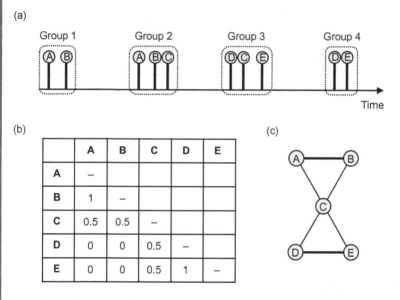

(b)

	A	B	C	D	E
A	–				
B	1	–			
C	0.5	0.5	–		
D	0	0	0.5	–	
E	0	0	0.5	1	–

Figure 9.2 Constructing a sociogram from RFID tag data: (a) small segment of a raw data stream of time-stamped individual observations at a feeder, with A–E representing five different birds; (b) the resulting sociomatrix showing association indices for each dyad using the simple ratio index; (c) the data from the sociomatrix represented as a sociogram.

> undirected sociomatrix shown in Figure 9.2b. The resulting sociogram is
> shown in Figure 9.2c. The edges in the sociogram are weighted according
> to the values of the SRI, with double-width lines linking dyads A–B and
> D–E. This is a toy example for illustrative purposes: far more data would
> be needed to construct a robust social network.

(SRI), which is the probability of observing both individuals together on a sampling occasion, given that at least one has been seen:

$$SRI = \frac{x}{x + y_{AB} + y_A + y_B}$$

where x is the number of occasions on which A and B were observed together, y_{AB} is the number of occasions on which A and B were observed but not together, y_A is the number of occasions on which just A was observed and y_B is the number of occasions on which just B was observed. This index varies between 0 (A and B never observed together) and 1 (always observed together).

The SRI is the appropriate metric if individuals are rarely missed, as is often the case for studies of captive populations. Other, more sophisticated indices of association are calibrated for the true probability of missing observations, but in the absence of the necessary calibration data, the SRI is the recommended metric. Free software is available for calculating association indices and simulating the effects of missing data [118].

9.3.2 Metrics from Social Networks

A social network, once constructed, can generate various metrics of sociality. These metrics are of two broad types: node level and network level. Node-level metrics are calculated for each node within a network. The most common node-level metrics are **degree** and **strength**, which are, respectively, the sum of the number of edges connected to the node of interest, and the sum of all the edge weights connected to the node of interest. Degree is appropriate for networks with binary edges, whereas strength is for networks with weighted edges. In the case of infectious disease transmission, for example, the degree might be the number of individuals with whom a focal individual has been in close proximity (say, within 2 m). An individual's degree would reflect their chances of being exposed to infection or infecting others.

Closeness centrality is derived from the shortest path lengths (number of edges) between the node of interest and all other nodes in the network. The shorter the paths connecting the node to other nodes, the higher its closeness centrality. This metric captures both direct and indirect associations and describes how well connected a node is to all other nodes in the network, and hence its ability to influence other nodes. In the case of disease transmission, closeness centrality captures the individual's potential to infect others in the network and the speed with which this may happen. A so-called **super-spreader** would be an individual with high degree or high closeness centrality. In the star network shown in Figure 9.1a, node A has high degree and high closeness centrality compared with the other nodes, whereas in the circular network in Figure 9.1b, all nodes have lower and equal degree and closeness centrality.

Betweenness centrality is the number of shortest paths between other nodes in the network that go through the node of interest. This metric captures a node's importance for connecting different parts of a social network. Node A in Figures 9.1a and d has high betweenness centrality.

Different node-level metrics will be highly correlated in some networks, while for other networks they may provide different information. For example, node A in Figure 9.1d has low degree compared with the other nodes but high betweenness centrality because it is the only connection between the two halves of the network.

Network-level metrics are calculated for the network as a whole, rather than for individual nodes, and reflect emergent properties of the associations within a network. **Density** is the number of edges present in a network divided by the total number of possible edges. Thus, a network with higher density has more edges per node. Figure 9.1c shows a network with maximum density. Density is a metric of the cohesion of a network. In the case of disease transmission, density is likely to affect the spread of disease within the network. **Cliquishness** describes the extent to which a network is divided into subgroups, or cliques, within which the nodes are highly connected. Figure 9.1d shows a network comprising two interconnected cliques. An infectious disease would be likely to spread quickly within cliques but more slowly throughout the network, compared with a network lacking this internal structure.

9.3.3 Hypothesis Testing

SNA is a valuable descriptive tool, but researchers may want to go further by testing specific hypotheses – for example, whether the pattern of associations observed in a network is different from that expected from random, whether males have fewer social interactions than females, or whether individuals associate according to a personality trait such as boldness or intelligence.

A key problem when testing hypotheses based on node-level metrics is that the measurements are not statistically independent. The nature of one individual's social interactions is clearly not independent of the interactions of the other individuals in the network. A common approach to addressing this problem of non-independence is to compare the metric of interest, calculated for the observed network, with the frequency distribution of the same metric calculated for each of a large number (e.g. 10,000) of randomly generated networks known as **null models**. The null hypothesis is rejected if the observed result is sufficiently rare in the distribution derived from the null models (e.g. in the 95th percentile or above). For example, if the difference between the mean degree for males and females is larger in the observed data than in 95 per cent of null models, the null hypothesis (that males and females do not differ in their number of social interactions) would be rejected. The tricky part is choosing the right method for generating the null models. Random networks can be generated by shuffling node labels, edges or observed group membership in the raw data. Which method is appropriate will depend on the nature of the data and the hypothesis being tested [119].

9.4 Dominance Hierarchies

In many social species, individuals engage in agonistic interactions in which one individual threatens, fights or supplants another when they compete for a valued resource such as food, shelter or a mate. Such agonistic interactions are rarely symmetrical, in that they usually end with one individual winning and the other losing the encounter. If the numbers of wins and losses are recorded for every individual in a group, it often becomes apparent that one dominant individual tends to win most of their interactions, whereas another subordinate individual loses most of theirs. In between the top- and bottom-ranking individuals are those who win some interactions but lose others. The overall arrangement of dominant and subordinate individuals in the group is referred to as a **dominance hierarchy.**

9.4.1 Constructing a Dominance Hierarchy

The basic dataset needed to construct a dominance hierarchy is a set of observations of the outcomes (i.e. wins and losses) of agonistic interactions between the majority of individuals in the study population. This information can be represented in two ways: as a two-column list containing the identity of the winner and loser of each sequential dominance interaction, or as a weighted directional sociomatrix that tabulates the wins and losses for all dyads (see Tables 9.2 and 9.3, respectively, in Box 9.2). The list format has the advantage of preserving information on the temporal sequence of interactions. A sociomatrix can be derived from a list of wins and losses, but the temporal information about the order of interactions is lost in the process, making it preferable to collect the raw data in the list format.

Once a dataset of dyadic interactions has been acquired, the next step is to use it to attempt to rank individuals from top to bottom. If every individual in the study population competes with every other individual an equal number of times, as might be the case in an organised sports tournament, individuals can be assigned a ranking based simply on their total number of wins (i.e. the row totals from the sociomatrix). However, if the individuals compete with each other different numbers of times, as is generally the case in behavioural data, simply counting total wins will not work. The method used to rank individuals must take account of the fact that some individuals have competed more often than others and consequently differ in the average calibre of their opponents. When assigning a rank, beating an individual who has won all their other contests should count for more than beating an individual who has won none.

Box 9.2 Deriving a dominance hierarchy using David's score

Here we illustrate the derivation of a dominance hierarchy from a sequence of observations of pairwise competitive interactions within a group of five subjects (A–E). An example might be a flock of starlings competing for access to a feeder. The raw dataset is a sequential list of 105 dominance interactions (Table 9.2), which is converted into a sociomatrix showing the observed numbers of wins and losses for each dyad (Table 9.3).

The David's score for an individual is calculated by weighting each dyadic success measure according to the unweighted estimate of the competitor's overall success, so that the relative calibre of the individual's competitors is taken into account (for a worked example, see [121]). In this case, the David's scores for individuals A–E are 4.4, −6.8, 7.4, 0.4 and –5.5, respectively. The resulting dominance hierarchy is therefore C > A > D > E > B.

Table 9.2 List of interactions (subset)

No.	Winner	Loser
1	A	D
2	C	D
3	C	B
4	E	B
↓	↓	↓
105	C	A

Table 9.3 Sociomatrix for dominance interactions

Winner	Loser				
	A	B	C	D	E
A	–	10	7	8	6
B	1	–	1	0	1
C	11	16	–	14	15
D	3	5	1	–	3
E	1	1	0	1	–

Various methods exist for inferring dominance hierarchies from unbalanced dyadic interaction datasets [120]. We will not attempt to review them all here, but in Box 9.2 we give an example using **David's score** – a method that takes a sociomatrix as input and which can be used to arrange individuals into an ordinal dominance hierarchy (i.e. first, second, third, etc.).

Most studies assume that dominance ranks remain stable over the period studied. If, however, the research question concerns the temporal stability of a dominance hierarchy, this can be addressed using information on the temporal sequence of interactions in list-format data. The **Elo-rating** method (originally developed to rank chess players) updates individual ratings after each single pairwise contest and can therefore be used to explore changes in rank over time.

9.4.2 Metrics from Dominance Hierarchies

The **linearity** of a dominance hierarchy is the extent to which individuals can be arranged into a single hierarchical order, from most dominant to most subordinate. Perfect linearity means that, for every dyad, one individual is dominant to the other, and that every triad is **transitive**; that is, for every three individuals A, B and C, if A dominates B and B dominates C, then A dominates C. In reality, dominance hierarchies are rarely perfectly linear: two or more individuals may have equal status (due to ties), and intransitive relationships may occur.

Landau's index of linearity (h) is a metric of the degree to which a dominance hierarchy is linear. The index is calculated as follows:

$$h = \frac{12}{N^3 - N} \cdot \sum_{i=1}^{N} \left(v_i - \frac{N-1}{2} \right)^2$$

where N is the number of individuals in the group and v_i is the number of individuals whom individual i has dominated. The index ranges from 0 to 1, with a value of 1 signifying perfect linearity. Values of h greater than 0.9 are generally taken to denote a strongly linear hierarchy.

A complementary metric is the **steepness** of a dominance hierarchy. Steepness reflects the size of the absolute differences between adjacently ranked individuals in their overall success in winning dominance encounters. Steepness is quantified as the absolute slope of the straight line fitted to individual dominance ratings (e.g. David's scores) plotted against the subjects' ordinal ranks [121]. Low steepness is characteristic of a relatively egalitarian group, whereas high steepness describes a more despotic group.

9.4.3 Reliability of Dominance Hierarchies

The minimum amount of data required to derive a reliable dominance hierarchy depends critically on the steepness of the hierarchy, with fewer observations being required for steeper hierarchies. A very steep hierarchy can be reliably inferred from as few as five interactions per individual. However, as the steepness cannot be known in advance, a conservative guide is at least 10–20 observed interactions per individual [120]. Thus, for a group of five subjects, at least 50–100 observed dominance interactions are needed to infer a reliable hierarchy. Many more observations would be required if the hierarchy is nearly flat.

Observations of dominance interactions commonly follow a skewed distribution, with many dyads being observed relatively few times and a few dyads observed many times. This is not a problem for inferring a dominance hierarchy unless there are many dyads for which there are very few observations.

A simple way of estimating the reliability of a dominance hierarchy, which can be used with any method for inferring dominance, is to split the dataset into two equal parts, calculate the dominance hierarchy for each half of the data, and then estimate the Spearman rank correlation between the ranks assigned to each individual in the two halves. If the correlation is high (>0.9), then the hierarchy is reliable and enough interactions have been collected [120].

Finally, we would urge caution when interpreting dominance hierarchies. A common error is to overgeneralise the meaning of a dominance hierarchy by treating the dominance status of each individual as though it were a fixed and general characteristic of that individual. In fact, dominance relationships are often fluid and capable of rapid change. Dominance relationships sometimes have a geographical element, with an individual's rank increasing towards the centre of its home range. Furthermore, a dominance hierarchy derived from one metric, such as competitive interactions over food, may not be the same as the hierarchy derived from a different metric, such as competition for roosting sites or mates [122].

9.5 Summary

- Social behaviour can be measured at different levels, from the behaviour of individuals to the behaviour of very large groups.
- The group is the basic unit of social organisation and must be clearly defined.
- It will often be important to measure group size. Crowding describes the average group size experienced by an individual.
- Individual identification is essential in many studies and can be accomplished either by artificially marking or tagging individuals, or by using natural variation. Marking and tagging have ethical and scientific implications.
- Social network analysis is the set of methods for describing and analysing how individuals interact within a group. Social network analysis yields metrics that describe properties of social interactions at both the individual and group levels.
- Dominance hierarchies rank the individuals within a group relative to one another and can be characterised in terms of their linearity, steepness and temporal stability.

10
Measurement Quality

Measuring behaviour, like measuring anything else, can be done well or badly. Poor-quality measurements lead to low statistical power and unrepeatable findings. Maximising the quality of measurements should therefore be a priority in the behavioural sciences.

Two basic issues must be considered when assessing the quality of behavioural measurements: **validity** and **reliability**. These terms are often confused in colloquial use, but they mean different things scientifically. The validity of a metric relates to whether the *right* quantity is being measured, whereas reliability relates to the extent to which measurements are *repeatable*. Validity and reliability are sometimes equated with **accuracy** and **precision** (Box 10.1). However, as we will explain below, while the accuracy and precision of measurements do affect their validity and reliability, there is no simple one-to-one mapping between these qualities.

Ideally, the measurements made in a study will be both valid and highly reliable. That said, validity is arguably more fundamental. As long as measurements are valid, lower levels of reliability can sometimes be tolerated. It is always better to measure the right quantity somewhat unreliably than to measure the wrong quantity very reliably. Hence, when evaluating the quality of behavioural data, validity is of primary importance. Although validity and reliability are theoretically independent of one another, they are often related in practice, especially in smaller datasets, where poor reliability can reduce the validity of measurements. Therefore, steps should be taken to make measurements as reliable as is feasible.

In this chapter, we define validity and reliability, describe how they are assessed and discuss what to do if they fall below acceptable levels. We also consider how floor and ceiling effects and outlying data points contribute to low-quality measurements.

Box 10.1 Accuracy and precision of measurements

Accuracy and precision are attributes of a set of repeated measurements of the *same* true quantity. Accuracy describes how close the mean of the measurements is to the true value of the quantity. Accurate measurements are free from systematic errors and are described as **unbiased**, whereas inaccurate measurements are biased.

An unbiased measurement consists of two parts: a systematic component, representing the true value of the quantity, and a random component arising from imperfections in the measurement process, known as **measurement error**. The smaller the measurement error, the more precise the measurements. The precision of a set of measurements describes the proximity of the individual measurements to each other. Precise measurements of the same quantity have low variance, whereas imprecise measurements have higher variance. While it can make sense to talk about the accuracy of a single measurement – how far the measurement is from the true quantity being measured – precision is always a property of a *set* of repeated measurements of the same true quantity.

For an infinitely large set of measurements of the same quantity, accuracy and precision are independent of one another. In practice, however, accuracy and precision are related because precision limits the accuracy of individual measurements. Even if the underlying measurement process is unbiased, small samples of imprecise measurements will tend to have inaccurate means as a result of chance. The lower the precision of measurements, the more measurements are required to obtain accurate estimates of a quantity.

If the true value of the quantity being measured is represented as the centre of a target, accurate shots will, on average, be centred on the bull's eye, whereas precise shots will be highly clustered in one area of the target (Figure 10.1).

A common scientific convention is to express the accuracy and precision of measurements in terms of the number of significant figures reported. The margin of error is generally taken to be half the value of the last significant place reported. For example, reporting a duration as 2.3 s would imply a margin of error of 0.05 s, whereas reporting 2 s would imply a margin of error of 0.5 s. In order to avoid accumulated rounding errors, it is good practice to retain more significant figures in the intermediate stages of a calculation than are finally reported.

Numerical results should not be reported in a way that implies **false precision** – that is, greater precision than is justified by the actual precision of the measurement method used. According to an old

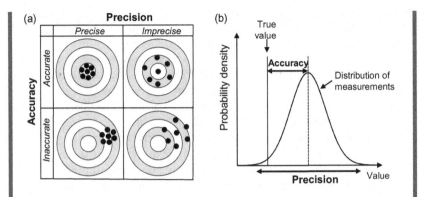

Figure 10.1 (a) Targets hit in four ways to illustrate the difference between accuracy and precision. (b) A graphical illustration of the statistical definitions of accuracy and precision.

scientific joke, a museum visitor asks a curator how old a dinosaur skeleton is. The curator replies that it is 60,000,005 years old. When asked how he knows this, the curator replies that a palaeontologist told him the skeleton was 60,000,000 years old when he started working in the museum 5 years previously.

10.1 Validity

Validity concerns the extent to which a metric actually measures what the researcher wants to measure – in other words, the *right* metric. A valid metric provides information that is scientifically relevant to the question being asked. Suppose, for example, that a researcher wants to know whether male rat pups spend more time suckling than female rat pups. In this instance, a valid metric might be the accurate measurement of the total duration of time that a pup is observed suckling in a session of continuous recording. The duration of suckling has strong **face validity** because the metric is clearly and directly related to the outcome variable specified in the research question.

The accuracy of measurements is an important component of validity, and inaccurate measurements will not be valid. However, validity means more than just accuracy. Unlike accuracy, validity is not an intrinsic property of a metric and cannot be assessed properly without reference to the scientific question. Returning to the above example, suppose instead that the researcher wants to know whether male rat pups receive more milk from their mothers than female pups. The exact same metric – suckling duration – might be chosen as a proxy for the amount of milk received because suckling behaviour can be quantified easily and accurately, whereas milk transfer is difficult to measure directly. However, suckling duration is only a *valid* metric if there is a strong positive correlation between the time spent suckling and the amount of milk received. In this case, the face validity of the metric is questionable. In some species, including rats and pigs, the relationship between suckling duration and milk intake is weak; rat pups and young piglets spend a lot of time suckling but obtain no milk for most of the time they suckle. Therefore, suckling duration is not a valid metric for milk transfer in these species, even though it can be measured accurately and reliably. Suckling duration would be described as having low **criterion validity** in pigs and rats because it does not correlate highly with the criterion (gold-standard) metric, which in this case would be the volume of milk received by a pup over the observed time period (a quantity that could be estimated by weighing rat pups before and after a known duration of suckling).

To take another example, suppose that a researcher wants to know whether male rat pups have a weaker mother–infant bond – a theoretically based **hypothetical construct** – than female rat pups. Suckling duration might again be chosen as the behavioural metric. Although this metric might in some respects reflect the strength of the mother–infant bond, it would probably suffer from low **construct validity** because it measures only one aspect of the complex interactions between pup and mother that are theorised to reflect the strength of the bond. Maternal licking behaviour and ultrasonic communications by the pup to the mother might also be central to measuring the mother–infant bond in rats. In this case, the metric does not fully capture the theoretical construct about which information is sought. Furthermore, even if suckling duration does correlate moderately well with the mother–infant bond, it may lack **specificity** as a metric, meaning that suckling duration could also correlate highly with other characteristics that are unrelated or weakly related to the mother–infant bond, such as the pup's body weight or aggressiveness. In this case, the metric would be said to lack **discriminant validity** because it is non-specific and fails to discriminate between several alternative phenomena.

In psychology and the social sciences, researchers commonly ask questions about so-called **latent variables**. These are variables that cannot be directly measured, either because this is practically difficult, as in the case of amount of milk transferred, or because the variable is a hypothetical construct, as in the case of the mother–infant bond. Latent variables must be inferred from other variables that *can* be observed and measured. In such cases, it is especially important to verify the validity of the metrics chosen. If invalid metrics are used and results are based on biased, irrelevant or non-specific measurements, the conclusions drawn from the data will be invalid, no matter how carefully other aspects of the study have been conducted.

10.1.1 Assessing Validity

Assessing whether a metric is valid involves considering the extent to which the measurements capture the phenomenon being studied and help to answer the question being asked. Valid measurements must be accurate – that is, unbiased estimates of the true quantity being measured. They must also be specific, meaning that they relate only to the quantity in question and nothing else. Most importantly, though, valid measurements must be of the *right* quantity.

Validity is assessed by comparing the metric with external sources of information about the phenomenon of interest. Typically, a gold standard or **ground truth** is required, against which the metric can be compared in order to judge whether it is measuring the right thing. How this is done will depend on the nature of the metric being validated and the extent to which it has been used previously in research. Widely used metrics are generally supported by substantial published evidence establishing their validity, whereas new metrics will require explicit validation.

Several different types of validation may be needed within a single study. For example, suppose a researcher wanted to discover whether birds exposed to unpredictable food supplies change their behaviour to conserve energy. The researcher might seek to test the specific prediction that birds with a history of being deprived of food for unpredictable periods of time spend a smaller proportion of the day flying compared with birds with a history of unlimited access to food. In this instance, the metric that requires validation is the proportion of the day flying: is it a valid metric of energy expenditure? To answer this question, several pieces of information are required. First, the researcher must establish whether birds that spend a greater proportion of their day flying actually expend more energy. This might appear obvious, but evidence is required nonetheless. At the very least, the researcher should cite published research that found higher metabolic rates in birds when flying. Ideally, there would be a study in which the time spent flying was manipulated experimentally in some way (e.g. by requiring birds to fly repeatedly between two perches to access food) and energy expenditure was measured directly (e.g. by calorimetry). Based on this sort of external evidence, the proportion of time spent flying could be validated as a behavioural proxy for total daily energy expenditure.

Behavioural metrics of emotional states such as anxiety, depression or pain are often validated by showing that they are changed in a predictable manner by drugs known to alter these states reliably in humans. Thus, the gold-standard approach for validating a behavioural metric of pain in a rat, such as facial grimacing, would be to show that the behaviour reduces in frequency when an analgesic drug is administered to rats that could reasonably be assumed to be experiencing pain (e.g. directly following surgery).

Even when the validity of a metric has been established in previously published research, it is still necessary to check the validity of the measurements in the current study because small differences in methodology could affect validity. The measuring instruments used in a study should be checked for accuracy if this is potentially an issue; for example, balances should be regularly zeroed and calibrated with a known mass.

When the coding of behaviour is automated, as described in Chapter 8, the automated method should be validated against manual coding of the same data – a process called **ground truthing**. For categorical data, this is usually done using a **confusion matrix** (or error matrix) to analyse the performance of an automated classifier (Box 10.2). A valid classifier will yield a high correlation between the outputs it produces and those obtained from manual scoring. For continuous data (e.g. the number of birds that are flying in a video frame), the **intra-class correlation coefficient** can be used to assess validity by comparing manual scores with automated scores (see Box 10.4).

When studying humans, researchers often use questionnaires to measure hypothetical constructs such as personality traits, subjective well-being, depression, intelligence or quality of life. While it might appear easy to

Box 10.2 Confusion matrices

For a behavioural category that is coded as either present or absent (e.g. in a video frame), a confusion matrix comprises a 2×2 table showing the number of occasions on which the automated method classified the behaviour as present or absent by the number of occasions on which the manual coding classified the behaviour as present or absent. In this case, manual coding is the ground truth.

Accuracy is calculated from the confusion matrix as:

$$\text{Accuracy} = \frac{\text{TP} + \text{TN}}{\text{TP} + \text{TN} + \text{FP} + \text{FN}}$$

This way of calculating accuracy will yield misleading results if the dataset is very unbalanced – that is, if the total number of positive and negative cases is very unequal, as in the example in Table 10.1, where the accuracy according to the above formula is 0.94 (where 0 indicates totally inaccurate predictions and 1 indicates totally accurate predictions). However, although prediction of negative cases is good (89 out of 90 correct), performance for the positive cases (5 out of 10 correct) is no better than chance.

The **Matthews correlation coefficient (MCC)** is a recommended alternative index that is unaffected by unbalanced datasets [123]. To get a high MCC score, the classifier must make correct classifications, both on the majority of negative cases and on the majority of positive cases, regardless of their ratios in the overall dataset.

$$\text{MCC} = \frac{\text{TP.TN} - \text{FP.FN}}{\sqrt{(\text{TP} + \text{FP}).(\text{TP} + \text{FN}).(\text{TN} + \text{FP}).(\text{TN} + \text{FN})}}$$

The MCC ranges between –1 (totally inaccurate) and +1 (totally accurate). It can easily be normalised to range between 0 and 1, where normalised MCC = (MCC+1)/2. The normalised MCC for the data in Table 10.1 is 0.81, reflecting the poor performance on rare positive cases.

Table 10.1 A confusion matrix showing example data

		Manual coding (ground truth)	
		Present (positive cases)	Absent (negative cases)
Automated classifier	**Present**	True positives (TP) 5	False positives (FP) 1
	Absent	False negatives (FN) 5	True negatives (TN) 89

devise such questionnaires, it requires expertise to construct valid instruments that actually measure what they are supposed to measure. The field of **psychometrics** is concerned largely with the construction and validation of questionnaires and rating scales. Psychometrics has spawned a number of concepts describing different aspects of validity and associated methods for assessing them. These include terms we have already mentioned. Criterion validity – how well the new metric relates to gold-standard metrics – is sometimes divided into **concurrent validity**, which describes how well the new metric correlates with established similar metrics measured at the same time, and **predictive validity**, which describes how well the new metric predicts the subsequent performance of subjects on a related criterion. For example, the predictive validity of a metric of cognitive ability in 5-year-olds might be assessed by the final level of education attained by the subjects many years later.

Perhaps counterintuitively, it can sometimes be advantageous for items in a questionnaire to lack face validity. Non-obvious questions can disguise the true aim of a questionnaire and make it less likely that respondents will manipulate their answers.

10.2 Reliability

Reliability is about repeatability – that is, the extent to which measurements are reproduced when a study is repeated under the same conditions. Mathematically, reliability is defined as the proportion of variance in a set of measurements that is due to the variance in the true quantity being measured (Box 10.3). Unreliable metrics have high measurement error, which means greater random variation in measurements. All else being equal, greater random variation results in reduced statistical power, and consequently unreliable measurements reduce the probability that a study will detect a true effect. Measurement error adds noise to data, making it harder to detect a true signal.

Many different factors affect how reliably behaviour is measured. Both the accuracy and precision with which individual measurements are made will influence the reliability of a dataset. The more accurate the constituent measurements, the smaller the measurement error. A dataset consisting of more accurate measurements will therefore be more reliable. Similarly, the

---

Box 10.3 Mathematical definition of reliability

The reliability index is calculated from the example shown in Table 10.2 as follows:

$$\text{Reliability index} = \frac{\text{true variance}}{\text{true variance} + \text{error variance}} = \frac{9.6}{9.6 + 12.8} = 0.43$$

A reliability index of 1 indicates that there is no measurement error and therefore measurements should be fully repeatable.

Table 10.2 Measurements from a sample of six subjects

Subject	Measurement	True value	Measurement error (= measurement – true value)
1	28	28	0
2	20	20	0
3	24	20	4
4	18	22	–4
5	26	22	4
6	16	20	–4
Variance of six values	22.4	9.6	12.8

---

greater the precision of measurements, the smaller the variance in the measurements and the smaller the measurement error. A dataset consisting of precise measurements will be more reliable.

10.2.1 Types of Reliability

Test–retest reliability describes the repeatability of measurements taken with the same methods and tools on the same set of subjects under the same conditions. The scientific instruments used in the measurement of behaviour, such as clocks, laser distance meters, location trackers and weighing balances, are potential sources of inaccuracy and imprecision leading to reduced reliability. Different brands or models of an instrument may yield metrics with differing reliability. The extent to which instruments have been properly calibrated, tested and maintained will also affect their ability to produce reliable measurements.

Measuring instruments are only one source of unreliability in behavioural research. A more fundamental problem is the involvement of human observers who operate the measuring instruments, administer rating scales or manually code the behaviour. The involvement of humans in the measurement process has a major bearing on the reliability of measurements because human behaviour varies within and between individual observers. The various effects of human observers therefore need to be considered.

Observers usually improve their ability to measure behaviour with practice and experience, which means that data from later recording sessions may be more reliable than data from earlier sessions, even if the written definitions remain unchanged. Observers may also suffer from fatigue and loss of concentration, so the reliability of data may decline towards the end of long recording sessions.

Intra-observer reliability describes the repeatability of measurements made by a single observer measuring the same set of subjects two or more times. **Inter-observer reliability** describes the repeatability of measurements made by two or more observers measuring the same set of subjects. Even if intra-observer changes in reliability are minimised, there may be stable individual differences in the reliability of data collected by different observers. For example, observers might apply subtly different criteria when measuring performance on a behavioural task, or subtly different interpretations of category definitions when scoring spontaneous behaviour. If several observers contribute data to a study, individual differences in reliability can reduce the overall reliability of a metric.

Some behaviour is intrinsically hard to measure reliably. Even with reliable instruments and reliable observers, some behavioural categories may be difficult to define clearly and recognise unambiguously. Some forms of behaviour occur very rapidly, making them hard to categorise accurately, and it can be unclear when a loosely defined category (e.g. play behaviour) starts and stops. A behavioural category that is not clearly and unambiguously defined probably cannot be recorded reliably.

10.2.2 Assessing Reliability

Reliability is assessed by investigating the repeatability of measurements across time and across different observers. This requires a sample of two or more *independent* replicates of measurements.

The dataset used in a reliability analysis is often a sample from the full dataset collected. This sample should be representative of the full dataset. The range of measurements in the reliability analysis must be similar to that in the main study because a truncated range in the sample used for the reliability analysis would result in an underestimate of reliability. The sample for reliability analysis should not be drawn from a single subject, observation session or video recording because the measurements would not then be independent of one another and could well be unrepresentative of the main study. Ideally, the sample for the reliability analysis should be chosen randomly and the measurements should be made under the same conditions as in the main study.

The choice of method for assessing reliability will depend on a number of factors, including the type of reliability estimate sought and the study design. Different statistics are required for categorical (nominal and ordinal) metrics and continuous (interval and ratio) metrics. Table 10.3 summarises some commonly used methods.

Cohen's kappa and its variants, Fleiss' kappa and weighted kappa, are ways of evaluating the reliability of two or more sets of categorical measurements. These statistics are more robust than simple percentage

Table 10.3 Common methods for assessing reliability (modified from [124])

Level of measurement	Number of replicate measurements between which reliability is to be assessed	Appropriate statistic
Categorical (nominal)	2	Cohen's kappa
	>2	Fleiss' kappa
Categorical (ordinal)	2	Weighted kappa
	>2	Fleiss' kappa
Continuous (interval or ratio)	≥2	Intra-class correlation coefficient

agreement because they take account of the agreement between two or more sets of categorical measurements that would be expected by chance, which is an issue when measurements can only take a small number of possible values and may therefore agree by chance. Cohen's kappa ranges in value between −1 and 1, with negative values indicating that the agreement between measurements is worse than would be expected by chance.

The **intra-class correlation coefficient (ICC)** is an index for evaluating the test–retest, intra-observer and inter-observer reliability of *two or more* sets of continuous measurements of the *same* metric. The ICC is distinct from the more familiar Pearson (or *inter*-class) correlation coefficient, which is used to measure the association between exactly *two* sets of measurements of *different* metrics. The ICC also differs from the Pearson correlation in that it reflects both the **consistency** (correlation) and the **absolute agreement** between two or more sets of measurements, whereas the Pearson correlation coefficient only reflects consistency (Figure 10.2). Two sets of measurements that have high absolute agreement will always be highly correlated, but it is possible to have high correlation with poor absolute agreement. The ICC can take values from 0 to 1, with 0 indicating totally unreliable measurements and 1 indicating totally reliable measurements.

Somewhat dauntingly, there are at least 10 different variants of the ICC, which are appropriate for different study designs and for answering different questions. Adding to the confusion, different authors and software packages recognise different subsets of these ICC variants and use different systems to classify them [125, 126]. Nonetheless, it is important to choose

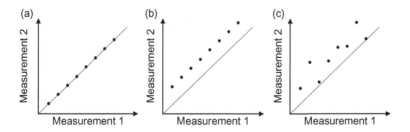

Figure 10.2 Scatterplots illustrating the difference between consistency and absolute agreement. (a) Perfect consistency and perfect absolute agreement. (b) Perfect consistency but lower absolute agreement. (c) Lower consistency and lower absolute agreement. The dotted lines indicate perfect consistency and agreement.

Box 10.4 Using the ICC to assess intra- and inter-observer reliability

Suppose a researcher wants to discover whether the duration of time engaged in stereotypic pacing by caged monkeys is associated with the size of cage in which they are currently housed (large or small). The plan is to measure pacing behaviour in videos of 20 monkeys housed in large cages and 20 monkeys housed in small cages. Owing to time constraints, the final data for analysis will be based on a single measurement made by a single observer of the proportion of time spent pacing by each monkey.

The researcher conducts an intra-observer reliability analysis to establish the reliability of their own measurements. They also conduct an inter-observer reliability analysis to establish whether their coding protocol is defined clearly enough to yield reliable measurements when used by other appropriately trained researchers.

The researcher chooses a sample of six monkeys from the main study for the reliability analyses: three monkeys are chosen randomly from each cage-size group to increase the representativeness of the sample. Videos of these subjects are coded twice by the researcher (observer 1), yielding two replicates of measurements (A and B) for intra-observer reliability analysis. The same videos are also coded once by each of two colleagues (observers 2 and 3), who have been trained on the measurement protocol, yielding two additional replicates of measurements that, together with one of the sets of measurements made by observer 1, can be used for inter-observer reliability analysis. The data collected is shown in Table 10.4.

Table 10.4 Illustration of data required to calculate intra- and inter-observer reliability

| Monkey | Columns used to assess intra-observer reliability | | | |
| | | Columns used to assess inter-observer reliability | | |
	Observer 1 measurement A	Observer 1 measurement B	Observer 2	Observer 3
1	0.51	0.58	0.65	0.43
2	0.29	0.22	0.34	0.23
3	0.33	0.35	0.49	0.35
4	0.44	0.59	0.59	0.53
5	0.02	0.01	0.11	0.25
6	0.81	0.89	0.95	0.71

For intra-observer reliability, the correct ICC is a two-way mixed-effects model (the observer is regarded as a fixed effect) with absolute agreement for measurements made by a single observer. For the data in Table 10.4, this yields an ICC of 0.96 (95 per cent confidence interval: 0.83–0.99), which could be described as 'good to excellent' [127].

For inter-observer reliability, the correct ICC is a two-way random-effects model with consistency for measurements made by a single observer (because this is what will be done in the full dataset). For the data in Table 10.4, this yields an ICC of 0.90 (95 per cent confidence interval: 0.64–0.98), which could be described as 'moderate to excellent' [127].

the correct ICC, and flowcharts are available to assist in this processes [127]. Box 10.4 gives examples of the ICC being used to measure intra- and inter-observer reliability in a common scenario.

How reliable must a behavioural metric be before it is deemed acceptable? For kappa coefficients and ICCs, there is no simple criterion above which reliability is acceptable and below which it is not. This is because factors other than reliability affect the value of these statistics, making their interpretation problematic. For example, both statistics will be higher if a larger sample of measurements is replicated in the reliability analysis. Despite this caveat, informal guidelines can be applied when judging reliability. For Cohen's kappa, values less than 0.40 are generally taken as signifying poor agreement, between 0.40 and 0.75 as fair to good agreement, and above 0.75 as excellent agreement. For ICCs, values less than 0.50 are taken as signifying poor reliability, between 0.50 and 0.75 as moderate reliability, between 0.75 and 0.90 as good reliability, and greater than 0.90 as excellent reliability.

The level of statistical significance (the p-value) of a reliability statistic says little about the degree of reliability because statistical significance depends on the sample size as well as the effect size. The value of the test statistic itself is what matters, not its statistical significance. An ICC of 0.50 would signify poor reliability but could still be highly statistically significant with a sufficiently large number of measurements. It is generally better to report the 95 per cent confidence interval for a reliability estimate rather than its p-value [127].

So far, we have focused on assessing reliability when the metric is based on measuring the quantity of interest only once in each subject. However, questionnaires used in human behavioural research often consist of multiple items (questions) that are all designed to measure the same underlying psychological construct. **Cronbach's alpha** is a statistic widely used in psychometrics to measure the internal consistency of the items that constitute a multi-item questionnaire or test [128]. Internal consistency will be high if all the items measure the same underlying construct. The individual items can be likened to independent observers, each producing an independent measurement of the same construct. Cronbach's alpha is a specific type of ICC in which data from different observers is replaced with data from the different constituent items [125]. Alpha ranges between 0 (no consistency between items) and 1 (perfect consistency between items). A low alpha implies that the items are not all measuring the same construct, and therefore calculating Cronbach's alpha is also part of validating a multi-item test. Unlike the other tests of reliability described above, for which perfect reliability is optimal, a score of ~0.9 is regarded as optimal for Cronbach's alpha. This is because Cronbach's alpha increases with the number of items, and a score of more than 0.9 suggests that the test contains too many items, or redundant items, and could therefore be made more efficient.

10.2.3 Dealing with Unreliable Measurements

In many cases, it is possible to improve the reliability of measurements if the assessment of reliability suggests that it is unacceptably low. Reliability can be improved in various ways.

Any measuring instrument used in a study should be properly maintained and have its accuracy and precision regularly checked. Choosing an instrument with a finer **resolution** of measurement can potentially improve reliability because the resolution of an instrument limits its accuracy and precision. All else being equal, a study in which the measurements are made with higher resolution will also be more reliable. For example, a measurement of the duration of a bird's call made using a clock that measures to the nearest second will, on average, be less accurate than one made using a clock that measures to the nearest millisecond: a call with a true duration of

5.26 s can, at best, be measured as 5 s using a clock that only measures to the nearest second. Measurements made with a lower-resolution instrument will also be less precise, on average, because large random errors in individual measurements will sometimes be exaggerated by the coarser measurement scale, adding to the variance in measurement. As a result of random measurement error, a call with a true duration of 5.26 s might be measured at 5.7 s using a clock that measures to the nearest tenth of a second, but this same measurement would be recorded as 6 s using a clock that only measures to the nearest second. However, the resolution with which an instrument measures only sets the *upper* limits for accuracy and precision. For example, a digital stopwatch may offer a resolution of hundredths of a second (0.01 s), but the human operator's reaction time of around 0.1 s means that repeated measurements of the same time interval will yield final digits (i.e. those representing hundredths of a second) that are effectively random because of the random error introduced by manual operation.

For some types of research, it helps to adopt **standard operating procedures (SOPs)**, whereby the protocol for collecting data is clearly specified. For example, SOPs have improved reliability in mouse behavioural phenotyping research, where it is essential for different laboratories to be consistent in how they run common behavioural tasks [129].

In observational research, where unreliability often stems from poorly defined behavioural categories, various things can be done to improve reliability. An unreliable metric can sometimes be redefined, or the recording method improved, to make it more reliable. Modifying the definition, perhaps by making it more restrictive in scope, may eliminate ambiguous cases that are difficult to categorise. If a category is difficult to record reliably using continuous recording, then time sampling may produce more reliable data. For some difficult categories, reliable measurement just requires a lot of practice combined with repeated reassessment of reliability until a pre-set criterion is reached.

Another approach is to combine two or more unreliable categories to produce a more reliable composite metric (see section 6.3). When the absolute frequencies of the separate scores are low and many individuals have scores of zero, the combined score may be more sensitive than any of its elements and easier to analyse. If metrics are combined in this way, the composite metric must have validity; that is, the separate metrics must all relate to the same entity and the composite metric must make scientific sense.

Where intra- or inter-observer reliability remains an issue, reliability can be improved by basing each data point on the mean of two or more measurements of the same quantity. These measurements can either be repeated measurements by the same observer or measurements made by two or more observers. This process is obviously more time consuming, but it may be the only way to improve reliability.

What if reliability cannot be improved? If statistical purity were the only arbiter, then all unreliable metrics would simply be discarded. This should certainly be the fate of any totally unreliable metrics, or of any moderately unreliable metrics that are of questionable relevance to a study's aim. However, a metric is sometimes so important that discarding it is not an option. If reliability cannot be improved, a universal strategy for mitigating the effect of moderately unreliable measurements is to increase the sample size – for example, by measuring more subjects. Increasing sample size improves statistical power and offsets the reduction in power arising from unreliable measurements.

10.3 Floor and Ceiling Effects

Validity and reliability are necessary attributes of high-quality measurements, but they are not always sufficient. Measurements may be both valid and highly reliable, yet still useless for answering a research question because of limited variation arising from **floor** or **ceiling effects**.

Two groups of subjects assigned to different experimental treatments may appear not to differ, when in reality they do, if all the measurements are clumped at one or other end of the possible range of values. Genuine differences will be obscured if all or most subjects obtain the minimum possible score (a floor effect) or the maximum possible score (a ceiling effect). For instance, a test of mathematical ability involving multiplication by two is unlikely to reveal differences between human adults because most people will answer all the questions correctly (a ceiling effect). A harder test is more likely to reveal differences, but a test that was too hard would result in most people answering none of the questions correctly (a floor effect). Although this issue may seem obvious, it is often overlooked as a possible explanation when negative results are obtained. Floor and ceiling effects apply to correlations as well as differences, as two variables will appear to be

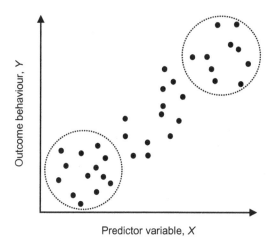

Figure 10.3 Scatterplot showing two variables that are highly correlated over their entire range. However, if the range of variation on the *x*-axis is restricted, the variables will appear uncorrelated, as illustrated by the random clouds of points within each of the dotted circles.

uncorrelated if either set of measurements is clumped at one end of its range (Figure 10.3).

Visual examination of pilot data should reveal whether floor or ceiling effects are likely. If they are, matters may be improved by choosing a different metric that produces a broader spread of measurements. Consider, for example, a choice test that measures which of two options is chosen by a subject when simultaneously presented. The test may show that one option is universally preferred by all subjects. However, a more sensitive measure of preference, such as latency to approach successively presented options, may reveal more variation between subjects in the attractiveness of the two options.

10.4 Outliers

An **outlier** is an apparently anomalous data point that lies outside the overall pattern of a distribution. Outliers sometimes result from gross errors in measurement or data entry, in which case they reduce the validity and reliability of measurements. It is therefore worth examining data for the

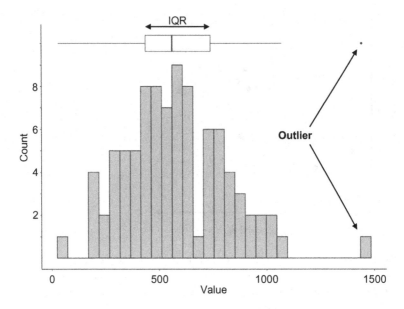

Figure 10.4 The distribution of the values comprising a metric, shown as both a histogram (below) and a box plot (above). In the box plot, the box is bounded by the first quartile (Q1) and the third quartile (Q3), with the median shown as a solid vertical bar. The left-hand whisker extends to the smallest data point greater than or equal to 1.5 times the interquartile range (IQR) below Q1. The right-hand whisker extends to the largest data point less than or equal to 1.5 times the IQR above Q3. The single outlying data point is shown as a small circle.

presence of outliers and considering whether it would be reasonable to remove them. Outliers can be detected visually by plotting the histogram or box plot of a metric (Figure 10.4).

The definition of what constitutes an outlier is arbitrary, but a common rule of thumb is that a data point may be regarded as an outlier if it falls more than 1.5 times the interquartile range (IQR) below the first quartile (Q1) or above the third quartile (Q3), where the IQR is defined as the difference between Q1 and Q3 of the data.

Outliers should be investigated carefully because they might represent true values, in which case a metric would be biased by removing them. To guard against this risk, outliers should *only ever* be removed if they can be proved to be due to an error or biologically impossible. For example, in sprinting, runners are justifiably disqualified if they move sooner than 0.1 s after the starting gun is fired because it is impossible for a human to react to a sound that quickly. If the removal of outliers is justifiable, the rationale for doing this and the mechanism for identifying outliers must be detailed in the study preregistration and analysis pipeline.

10.5 Summary

- Poor-quality measurements are likely to yield meaningless or unrepeatable findings.
- High-quality measurements are characterised by validity and reliability.
- Validity relates to whether the right quantity is measured and is assessed by comparing a metric with a gold-standard metric.
- Reliability relates to whether measurements are repeatable and is assessed by comparing repeated measurements.
- The accuracy and precision with which measurements are made affect both validity and reliability.
- A major source of unreliability in behavioural data comes from the involvement of human observers in the measurement process.
- Where trade-offs are necessary, it is better to measure the right quantity somewhat unreliably than to measure the wrong quantity very reliably.
- Floor and ceiling effects can make measurements useless for answering a question, even if they are valid and reliable.
- Outlying data points should only be removed if they can be proved to be biologically impossible or to result from errors.

11
Data Analysis

11.1 Behaviour Needs Statistics

The pioneering physicist Ernest Rutherford once said: 'If your experiment requires statistics, you ought to have done a better experiment.' While this may once have been true in some areas of physics, it is not true of most behavioural science. Behaviour is complicated and variable, and researchers usually do need statistics to detect patterns in data and test hypotheses.

Statistical analysis sometimes strikes fear into students, but knowing how to analyse data is an empowering and satisfying skill. Statistical analysis is best thought of as a set of tools. If used properly, statistics help researchers to describe datasets and answer questions with them. But like any tool, statistics can be misused or even intentionally abused, so it is important to use them wisely and recognise common mistakes and abuses. A basic understanding of statistics is essential for reading scientific publications critically and integral to good research design. Whether or not you need to analyse data yourself, it is simply not possible to be a good behavioural scientist without at least a basic grasp of statistics.

This is not a statistics textbook and there is far too little space here to provide a full prescription for statistical analysis. Instead, we aim to provide broad guidelines for navigating the bewildering array of statistical methods available. To do this, we need to introduce a bit of theory. For a full introduction to statistical methods, there are many excellent books on this subject (e.g. [130–134]).

Approaches to statistical analysis differ between scientific disciplines. These differences often obscure the existence of common general principles that underlie apparently different statistical methods. One of our aims in this chapter is to show that the plethora of tests commonly taught in introductory statistics courses can be thought of as variants within a common framework. Once this framework is understood, choosing the correct statistical method becomes less daunting.

This chapter describes only the tip of an iceberg of available methods. We have focused on the methods that we consider most generally important. There is no substitute for reading recent primary research publications to understand what methods researchers are currently using in a specific area.

11.2 General Advice on Statistics

Data analysis should always be considered when initially designing a study. Failure to do this can result in costly mistakes of collecting unanalysable or uninterpretable data. A detailed analysis plan should form part of any study proposal or preregistration and is required for power analysis (see section 4.9). This plan should identify the outcome and predictor variables to be used, whether any data points will be excluded and exactly how the data will be analysed. The plan should be as specific as possible in order to reduce researcher degrees of freedom and avoid p-hacking (see Chapter 2).

Consulting a statistician *before* starting to collect data is a wise precaution. That said, working with a statistician should not be an excuse for opting out of understanding statistics. Statistical analysis is too central to most studies to risk transferring responsibility to someone who might not fully appreciate the aims of the study.

It is important not to become carried away with statistics and perform analyses that are more complicated than is necessary to answer the research question. Simple statistics can often be very powerful and have the advantage of being easier to interpret and communicate. The trick is knowing when a more complex approach is justified.

Fancy statistics cannot fully compensate for a poorly designed study or unreliable measurements. For example, if there is reason to believe that the outcome variable might differ according to the sex of the subject, but the effects of sex are not the main focus of the study, this issue can be dealt with at the design stage by including equal numbers of male and female subjects in each of the treatments of interest. With a **balanced design** of this type, the effects of sex can justifiably be ignored at the analysis stage because sex cannot bias the results. With an unbalanced dataset, it may be possible to control for the effects of sex statistically by including it as an additional predictor variable in the analysis, but this can become problematic if sex is correlated with treatment. In general, the better a study is designed, the simpler the analysis can be. This is true for both experimental and observational studies.

Choosing which software package to use for statistical analysis is an important decision. Becoming proficient in any package takes time, so it is worth considering whether a package meets both current and probable future needs. SPSS and the statistical programming language R are currently the two packages most commonly used by behavioural scientists. SPSS is an expensive proprietary package, whereas R is open source and free. Numerous statistics textbooks are based around each of these packages, and many resources are available online. SPSS has a point-and-click interface, making it easier to use initially; R requires the user to write code, making it slower to learn and often frustrating for the beginner. Unlike SPSS, for any analysis beyond the most basic, R forces the user to write a saveable **script** to implement an analysis. A script has the advantage that it can be shared with students, collaborators and reviewers and published alongside a study, providing transparency and facilitating reproducibility. Further advantages of R include the facility to produce customisable publication-quality figures and the growing number of contributed R packages that implement specialist methods. Finally, R is a programming language, making it relatively straightforward to incorporate complex data manipulation, automation of repetitive procedures and simulations into an analysis pipeline.

The steps in a standard data analysis pipeline are shown in Box 11.1. In the following sections, we describe each of these steps.

11.3 Data Preparation

Before data can be analysed, it must be formatted correctly for importing into a statistical package. Data will usually be saved in a spreadsheet. Column headings should be put in the very first row of the spreadsheet and should only occupy a single row. Headings should be short but descriptive and should not include spaces or symbols, as these can confuse statistics packages.

For many applications, it is easier if the data table has the correct shape – specifically, a long format known as **tidy data** [135]. In a tidy data table, each variable is one column and each observation is one row (see Tables 11.1 and 11.2 for examples of untidy and tidy data, respectively). Data on the same metric should be collated in the same table. However,

> **Box 11.1** Steps in a standard data analysis pipeline
>
> 1. Prepare data tables in a spreadsheet package.
> 2. Save a definitive version of the dataset.
> 3. Import the data into a statistical analysis package.
> 4. Assign correct data types to columns.
> 5. Calculate descriptive statistics.
> 6. Examine outlying data points and correct data entry errors.
> 7. Visualise the data in the form of histograms, boxplots and scatterplots.
> 8. Conduct pre-specified confirmatory analysis to answer research questions.
> 9. Check model assumptions are valid and, if not, make the necessary adjustments to analysis.
> 10. Calculate effect sizes and their confidence intervals, as well as p-values.
> 11. Consider multiverse analysis to explore the robustness of findings to analytic decisions.
> 12. Conduct exploratory analysis and generate new research questions.

Table 11.1 An example of untidy data with two observations of 'measurement' per row

Subject	Control	Experimental
Tom	5	10
Dick	3	12
Harry	7	14

different metrics may need to be placed in separate tables. Additional columns are required to identify each observation in a tidy data table. For example, column headings for an experiment in which repeated measurements are made on subjects exposed to different treatments might be: 'Subject', 'Treatment', 'Date', 'Time' and 'Measurement'.

Tidy data is uneconomical with space because of the additional columns required to identify each measurement, but the format facilitates data manipulation, visualisation and modelling. Most statistics packages contain functions for reshaping data tables, so it is not a disaster if a dataset has been collected in a more economical untidy format.

Table 11.2 The same data as in Table 11.1 reformatted as tidy data with one observation per row

Subject	Treatment	Measurement
Tom	Control	5
Dick	Control	3
Harry	Control	7
Tom	Experimental	10
Dick	Experimental	12
Harry	Experimental	14

Each column in a tidy data table must be assigned a data type. Details of what is required will depend on the package, but it will generally be necessary to decide the scale or type of data in each column (e.g. nominal/ordinal/interval or character/integer/numeric), and for columns that contain categorical predictor variables, it may be necessary to assign these as factors. For example, in Table 11.2, 'subject' would be a factor with three levels (Tom, Dick and Harry), 'treatment' would be a factor with two levels (control and experimental) and 'measurement' would be an integer.

The basic data table should be kept as simple as possible. Any calculations whereby one or more variables are used to derive a new variable (e.g. rates or proportions, or more complex transformations of variables) should be part of the statistical analysis pipeline to provide transparency over how derived variables have been calculated.

The final stage of data preparation should be to archive a definitive version of the raw dataset somewhere safe. This dataset should not be changed after data analysis is started. If errors identified later during subsequent analysis need to be corrected, this should be shown as part of the documented analysis pipeline.

11.4 Describing Data

11.4.1 Populations and Samples

The distinction between populations and samples is critical to statistical analysis. The **population** is the entire set of items or individuals (e.g.

animals, vocalisations, testosterone levels, humans or adult male rhesus macaques) under consideration and about which inferences are to be made. The population is distinguished from a **sample**, which is the particular subset of entities selected from the population for a study. Measurements are made on a sample – for example, 30 macaques chosen randomly from a laboratory colony of 100 – and these measurements are used to draw statistical inferences about the population as a whole – for example, the whole colony, all macaques living under comparable conditions or members of all closely related species. In some rare cases, it may be possible to measure the whole population of interest, but this is atypical, and the methods described in this chapter assume that measurements are made on a sample.

11.4.2 Descriptive Statistics

Descriptive statistics include measures of central tendency (e.g. mean, median, mode), variation (e.g. standard deviation, IQR) and other descriptors of distributions, such as skew (a measure of symmetry) and kurtosis (the extent to which data points are clustered near the peak or tails relative to a normal distribution).

In the unlikely event that an entire population is measured, the descriptive statistics, such as the mean and standard deviation, are called **parameters**. The parameters are the true values required to test hypotheses definitively. Consider, for example, the question of whether male macaques are more aggressive than females. If the frequency of aggression were measured in the entire population of macaques, and the mean frequency of aggressive behaviour was higher in males than females, then the question would have been answered: males are on average more aggressive than females in this population. No further statistical tests would be required. However, as we have said, very few studies measure entire populations.

In the much more typical case where only a sample of macaques has been measured, the parameters of interest cannot be calculated directly but instead must be estimated from the sample. The methods for estimating parameters and for drawing conclusions from these estimates are known as **inferential statistics**.

Before embarking on inferential statistics, it is always a good idea to get to know the data. Calculating descriptive statistics for the sample is the first step. There are several good reasons for calculating descriptive statistics. For a start, they are useful for revealing data entry errors and outliers (see section 10.4). For instance, if descriptive statistics reveal that, contrary to expectations, male macaques have a lower mean body mass than females, it would be worth checking that the subjects' sex has been coded correctly. (To minimise coding errors, it is generally better to use descriptive codes such as 'M' and 'F' rather than, say, 1 and 0.) If an unexpected pattern turns out to be real, it is useful to know that the sample may be unrepresentative in some respects. Descriptive statistics can also inform the correct choice of inferential statistics. If a histogram reveals extreme skew, a lot of zeros or data bounded between two values (as might be the case for proportions), this should raise a red flag. Such variables are likely to require special treatment – for example, transformation (section 11.6.1), the use of generalised linear models (section 11.8) or possibly non-parametric methods (section 11.8.1). If the dataset includes several continuous variables, it may be useful to construct a **correlation matrix** to explore how each pair of variables is related. When two variables are correlated at more than 0.7, they should not both be included as predictor variables in the same statistical model because they carry much of the same information. Finally, by revealing unexpected patterns in the data, descriptive statistics can inspire exploratory analysis (section 11.12).

11.4.3 Data Visualisation

Descriptive statistics alone are rarely sufficient to understand a dataset. Some form of visualisation nearly always helps. Histograms are often easier to interpret than tables of descriptive statistics and should always be plotted for continuous variables. To understand how two variables are related to one another, there is no substitute for a scatterplot.

Anscombe's quartet is a famous set of four fictitious datasets that demonstrates the value of plotting data (Figure 11.1) [136]. The four datasets yield *identical* descriptive statistics, and identical regression coefficients, but reveal completely different patterns when plotted. Linear regression would be an appropriate analysis for dataset (a), but dataset (b) is clearly non-linear and should be fitted with a curve, while datasets (c) and (d) have single outlying points with high leverage (i.e. a big effect on the

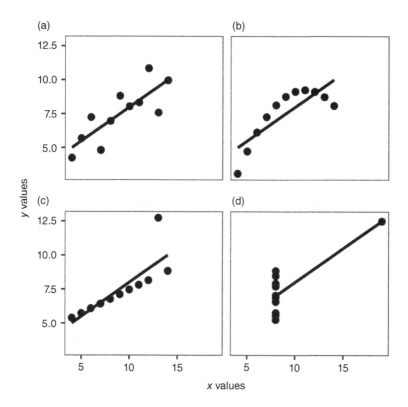

Figure 11.1 Anscombe's quartet. Scatterplots of four datasets, (a)–(d), each showing the best-fitting linear regression line. The following statistics are identical for all four datasets: number of points = 11, mean of x values = 9.0, mean of y values = 7.5, equation of regression line: $y = 0.5x + 3.0$, regression coefficient = 0.5, standard error of regression coefficient = 0.118, $R^2 = 0.667$.

fitted line). Indeed, a model with a continuous x variable looks totally inappropriate for dataset (d).

11.5 Using Data to Answer Questions

Inferential statistics are methods for making inferences about the population from which a sample was drawn. These methods involve two steps: first, using the data to estimate parameters of the population (section 11.4.2), and second, using these **parameter estimates** to test hypotheses about the population.

The dominant approach to testing hypotheses is **null hypothesis significance testing (NHST)**, whereby the parameter estimates are used to estimate the probability (*p*-value) of observing an effect at least as extreme as that actually observed in the sample, given that the null hypothesis is true. If the *p*-value is below a pre-set critical value, usually 0.05, the null hypothesis is rejected, whereas if the *p*-value is above the critical value, the null hypothesis is retained.

The traditional approach to presenting and interpreting results from inferential statistics has been to focus on the binary outcome of NHST, with undue importance attached to whether or not a test is 'statistically significant' (i.e. *p*-value ≤ 0.05). This approach is misguided, for a number of reasons. The *p*-value is a continuous measure and 0.05 is an arbitrary criterion for deciding whether a test is significant or not. More importantly, the *p*-value provides no information about the *size* of the effect detected. All else being equal, the *p*-value will get smaller as the sample size gets larger. Consequently, tiny effects with little biological, medical or societal significance can be highly statistically significant, provided a large enough sample is measured. With a sample size of 100, for example, a correlation of 0.2 is statistically significant, even though a correlation of this size represents a very weak association ($R^2 = 0.04$, meaning that only 4 per cent of the variation in one variable is accounted for statistically by variation in the other).

In contrast, estimation statistics, which focus on parameter estimates and their confidence intervals, provide much more useful information because they describe the size of the effects and the precision of the estimates. The term **effect size** is used for any statistic that quantifies the *degree* to which the sample results diverge from the null hypothesis (e.g. no difference in group means or no association between two variables; see Box 2.1). Effect sizes are continuous measures calculated from parameter estimates. Examples include the standardised mean difference, the correlation coefficient and R^2.

One response to the replication crisis has been a shift away from NHST to focusing more on estimation statistics. When comparing the findings from different studies, it is more informative to compare effect sizes than to compare statistical significance because differences in significance between studies can arise simply as a consequence of differences in sample size, whereas effect sizes are independent of sample size.

11.5.1 Statistical Models

Emphasising parameter estimation over NHST goes hand in hand with regarding inferential statistics primarily as ways of *building models* rather than *conducting tests*. Generally speaking, models are simplified representations of the world. In the case of statistical analysis, models are representations of a hypothesis about the true patterns present in the population from which the sample is drawn. For example, the hypothesis that male macaques are more aggressive than females assumes a model of the population in which mean aggression is higher in males than in females; the hypothesis that higher blood testosterone predicts higher aggression in macaques assumes a model in which there is a positive association between testosterone and aggression. Although statistical inference always involves specifying and fitting a model to the data, this principle is not always made explicit or taught.

The basic aim of statistical models is to understand how the outcome variable changes as the value of a predictor variable (or variables) changes. A statistical model can be represented graphically or as a mathematical equation. The outcome variable, which is usually the main variable measured in a study, is designated the y variable and is represented on the vertical axis of graphs or on the left-hand side of the model equation. The predictor variables are designated as x variables and are represented on the horizontal axis of graphs or on the right-hand side of the model equation. In the macaque example above, the metric of aggression is the y variable and the x variable is sex (a categorical variable with two levels) or blood testosterone (a continuous variable measured on a ratio scale). Figure 11.2 shows graphical representations of two models of aggression in macaques. In model (a), the effect of sex on aggression is shown by the difference between mean aggression in males and females, and in model (b) the effect of testosterone on aggression is shown by the slope of the best-fitting regression line. The precision of the estimate of each effect is shown by the 95 per cent confidence intervals for the difference in means and for the slope, respectively.

The graphical approach works well for relatively simple models and can be more intuitive for the less mathematically minded. However, illustrating models graphically becomes more difficult with complex models involving multiple predictor variables. Therefore, statisticians often prefer to express

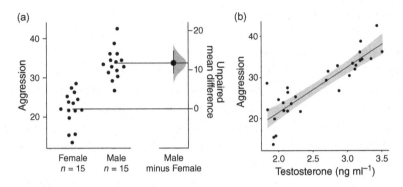

Figure 11.2 A fictitious dataset comprising measurements of an aggression score from 30 macaques (15 of each sex) showing two alternative models. In model (a), aggression is predicted by sex. The graph shows a Gardner–Altman plot: the individual data points are shown on the left; the horizontal lines represent the mean aggression for each sex; and the difference between the means (an unstandardised measure of effect size) is shown on a separate axis on the right. The 95 per cent confidence interval is shown by the vertical lines above and below the point estimate of the mean difference. In model (b), aggression is predicted by testosterone. The graph shows a scatterplot of the 30 data points fitted with a linear regression line (95 per cent confidence interval shown in grey). In both (a) and (b), the lines are the graphical representations of the statistical model fitted.

their models in terms of equations. Simplified verbal equations can be used to replace the mathematics. For example, the verbal equations corresponding to models (a) and (b) in Figure 11.2 are:

Aggression ~ Sex

and

Aggression ~ Testosterone

In these equations, the tilde (~) replaces the equals sign and is read as 'is predicted by'.

11.6 General Linear Models

The general linear model (GLM) is the primary workhorse of inferential statistics. It is not a single model but a family of related models that should

Table 11.3 Common statistical tests and their GLM equivalents

Statistical test	Verbal equation for GLM equivalent
Two-sample *t*-test (assuming equal variance)	$y \sim x_1$, where x_1 is a categorical variable with two levels
One-way ANOVA with three groups	$y \sim x_1$, where x_1 is a categorical variable with three levels
2 × 2 factorial ANOVA	$y \sim x_1 + x_2 + x_1 \times x_2$, where x_1 and x_2 are both categorical variables with two levels each
Linear regression	$y \sim x_1$, where x_1 is a continuous variable
Multiple regression	$y \sim x_1 + x_2$, where both x_1 and x_2 are continuous variables
ANCOVA	$y \sim x_1 + x_2$, where x_1 is a categorical variable and x_2 is a continuous variable

ANOVA, analysis of variance; ANCOVA, analysis of covariance.

be considered whenever the outcome variable is measured on a continuous scale. The GLM provides a flexible framework for modelling how one or more predictor variables affect a continuous outcome variable [137]. In a GLM, the predictor variables can be categorical, continuous or a mixture of the two. The two models in Figure 11.2 are both examples of simple GLMs. Model (a) is the equivalent of a two-sample *t*-test and model (b) is the equivalent of linear regression. These examples illustrate how the GLM family subsumes many of the separate statistical tests designed for specific configurations of predictor variables. Further examples are given in Table 11.3, but a vast number of other specifications are also possible.

If more than one predictor variable is required in a GLM, then additional *x* variables are simply added to the right-hand side of the equation. For example, the model

Aggression ~ *Sex* + *Testosterone*

states that aggression is predicted by both the sex of a macaque and its testosterone (Figure 11.3a). This model tests the hypothesis that there is an effect of sex on aggression once the testosterone of an individual is known, and conversely that there is an effect of testosterone once sex is known. The GLM in this case assumes that the effects of sex and testosterone are **additive**; that is, the effect of one predictor variable is independent of the value of another predictor variable. In Figure 11.3a, additivity means that

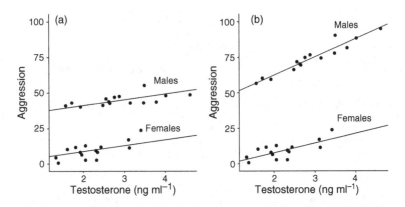

Figure 11.3 Two more fictitious datasets in which aggression is predicted by both sex and testosterone. In dataset (a), the effects of sex and testosterone are additive, whereas in dataset (b) there is a non-additive interaction between sex and testosterone.

the fitted lines for males and females are parallel, showing that the difference in aggression between males and females is the same whatever the value of testosterone.

The effects of predictor variables may not be additive, however. An **interaction** is said to be present if the effect of one predictor variable varies according to the value of another predictor variable. Interactions between any of the predictor variables in a GLM can be included in the model to capture such non-additive effects. In the macaque example, the GLM can be modified to accommodate a non-additive interaction by the addition of an additional term to the model equation:

Aggression ~ Sex + Testosterone + Sex × Testosterone

This model is illustrated in Figure 11.3b, in which the fitted lines for males and females are no longer parallel, showing that the difference in aggression between males and females is larger at higher values of testosterone.

Despite its name, the GLM is not restricted to modelling linear associations between continuous variables. Data requiring a curve, such as that shown in Figure 11.1b, can be modelled by adding a squared term to the model equation:

$y \sim x + x^2$

By extension, more complex curves can be modelled by adding further polynomial terms as required (x^3, x^4, etc.).

11.6.1 Assumptions of the GLM

The GLM relies on a number of assumptions. Parameter estimates are liable to be inaccurate and inferences from models invalid if these assumptions are violated.

The major assumptions of the GLM concern the **residuals**, which are the differences between individual data points and their associated **fitted values** (i.e. the values predicted by the fitted model). The GLM assumes that the residuals are normally distributed with a mean of zero (the assumption of **normality**), that they have constant variance at all values of x (the assumption of **homoscedasticity** or **homogeneity of variance**) and that they are independent of one another. Some texts state that the raw data must be normally distributed for a GLM to be valid, but this is incorrect; it is the *residuals* that must be normally distributed.

In practice, the assumptions of normality and homogeneity of variance are commonly violated in behavioural data. The precision with which measurements are made is often proportional to the magnitude of the quantity being measured, which means that small quantities are measured absolutely more precisely than large quantities. This has the effect of creating residuals for which the distribution is positively skewed and the variance increases as the value of x increases. Metrics expressed as proportions are by definition bounded between 0 and 1, which means that residuals are unlikely to be normally distributed, particularly if many data points are near to the extremes of the distribution.

Small deviations from normality are unlikely to cause major problems with the GLM, and parameter estimates will be unbiased even if the normality assumption is violated. Therefore, as long as the distribution of residuals is symmetrical and highest in the middle, it need not be perfectly normal. Violations of homogeneity of variance should be taken more seriously.

Both normality and homogeneity of variance can often be improved by applying an appropriate **transformation** to the outcome variable before fitting the model. Table 11.4 gives some useful common transformations. If

Table 11.4 Common transformations to improve normality and homogeneity of variance

Name	Transformation	Recommended use	Notes
Square root	$y \rightarrow \sqrt{y}$	Often used for count data, especially if values are small. The mildest transformation for reducing right skew and homogenising variance.	Add a constant before square-rooting if there are negative values in the dataset.
Log	$y \rightarrow \log(y)$	Often used for measurement data. A stronger transformation for reducing right skew and homogenising variance.	It does not matter whether natural logs or base-10 logs are used. Add a constant before logging if there are negative values or zeros in the dataset.
Reciprocal	$y \rightarrow 1/y$	The strongest transformation for reducing right skew and homogenising variance.	Add a constant before taking the reciprocal if there are negative values or zeros in the dataset. Note that the reciprocal reverses the order of data points with the same sign.
Logit	$y \rightarrow \log\left(\frac{y}{1-y}\right)$	Often used for continuous (i.e. non-count-based) proportions. For count-based proportions, use GLiM [138].	y variable must be a proportion between 0 and 1. Not usually necessary if all values are in the range $0.3 < y < 0.7$.

a transformed outcome variable is used for statistical analysis, then any results, such as graphs, descriptive statistics and parameter estimates, should be back-transformed for presentation because transformed values are hard to interpret. In cases where normality and homogeneity of variance cannot be corrected with a transformation, it will be necessary to use **generalised linear models (GLiMs)** (section 11.8), or resort to non-parametric statistics (section 11.8.1).

The assumption of independence is also commonly violated in behavioural data. Datasets are often structured such that groups of data points are likely to be more similar to each other than they are to other groups of data points. Most obviously, if multiple data points are collected from the same subject, they are likely to be more similar to one another than to data from other subjects. Similarly, data points from the same family, territory, nest or school are likely to be non-independent because of shared genetic or environmental factors. Data points collected by the same observer, on the same day or under similar weather conditions may also be similar. If data points in a group are similar, they will have similar residuals, violating the assumption of independence.

Non-independence affects the power of statistical models to detect true effects. Power may either be increased or decreased, depending on the circumstances. Consider, for example, the case of multiple measurements made on the same subjects. If the predictor variable of interest is varied within subjects, as occurs in a paired or within-subjects design, a GLM that does not include this information will typically be less powerful than one that does. In contrast, if each subject occurs in only one level of the predictor variable, but multiple measurements are made on each subject, a GLM that does not include this information will be more powerful than one that does because the relevant sample size is inflated by the repeated measurements of the same individual (so-called **pseudoreplication**).

Non-independence can be eliminated from statistical models in two ways. Some sources of non-independence can be accounted for by using tests designed for repeated measurements of the same subjects: for example, two-sample *t*-tests are replaced with paired *t*-tests and ANOVAs (analysis of variance) with repeated-measures ANOVAs. Pseudoreplication can be eliminated by using the means of each related group of data points (subject, family, school, etc.). However, this approach is unsatisfactory. Taking a mean is arbitrary; why not calculate a median instead? Moreover, taking any measure of central tendency is wasteful because it throws away useful information about the variation within each group of data points. If the variation among the repeated measurements from a subject is small, it is possible to make a more precise estimate of the true value for that subject than if the variation is large. Taking means also does not allow for multiple nested levels of non-independence, such as a dataset in which there are not only multiple measurements from each subject but also multiple subjects from each family, territory, cage or school. A better approach to tackling non-independence is to use all of the data and explicitly account for any suspected sources of non-independence by using a **mixed-effects model**.

11.7 Mixed-Effects Models

General linear mixed-effects models (GLMMs; also known as random-effects models and multi-level models) are extensions of the GLM family. They provide a widely used method for dealing with sources of non-independence arising from spatial or temporal autocorrelation between data points.

Perhaps the most common reason for using mixed models in a behavioural study is to deal with non-independence arising from repeated measurements of the same subjects. For example, the starling data described in Box 4.1 comprised 20 measurements of vigilance under high predation risk and 20 under low predation risk. However, these were not 40 otherwise-independent measurements: rather, they were 20 pairs of data points from 20 subjects, each measured under high and low risk (Figure 11.4a). Data of

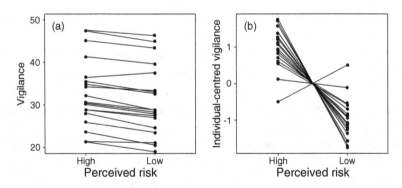

Figure 11.4 (a) Paired vigilance data from 20 starlings measured under two levels of predation risk. (b) The same data plotted with each individual's data expressed relative to its mean vigilance; zero on the *y*-axis represents the mean vigilance for each bird. This transformation removes between-individual variation in vigilance so that the effect of risk can be seen more clearly. Including a random intercept for bird in the GLMM has an analogous effect and hence increases the power of the model to detect an effect of risk.

this type can be analysed with a paired *t*-test or, if there were more than two levels of predation risk, a repeated-measures ANOVA. These tests are equivalent to specific cases of the GLMM in which an additional predictor – in this case 'subject' – is added to the model to account for the fact that the dataset contains repeated measurements on individual birds.

The GLMM is so called because it contains a mixture of two types of predictor variables: **fixed effects** and **random effects**. The fixed effects are the predictor variables that are present in a standard GLM. They are the variables whose effects the researcher wants to estimate. In contrast, the random effects are additional predictor variables added to the model to account for sources of non-independence in the data. Random effects are always categorical variables, and there is generally no interest in estimating their effects. The levels of a random effect are assumed to be a random sample of levels drawn from the population of all possible levels. In the case of the starlings, the assumption is that the 20 birds are a random sample of all possible starlings.

The distinction between fixed and random effects can be hard to grasp at first. A helpful way of deciding whether a variable is a fixed or a random effect is to ask two questions. First, is the aim of the study to make specific inferences about the levels of the predictor variable, or general inferences about the population from which the levels are drawn? If the answer is specific inferences, then the predictor is a fixed effect; if the answer is general inferences about the population, then the predictor is a random effect. Second, if someone were to repeat the study, would they use the exact same levels of the variable used in the original study? If the answer is yes, then the predictor in question is likely to be a fixed effect, whereas if it is no, then the predictor is likely to be a random effect. In the starling example, the interest is in how starlings in general respond to specific levels of predation risk. A replication study would attempt to copy the levels of predation risk exactly but would perform the study on a new randomly chosen sample of 20 birds. The answers to both questions specified above indicate that predation risk is a fixed effect and subject is a random effect.

Subject will almost always be a random effect, whereas other variables can be either fixed or random, depending on the study. For example, if the same experiment is conducted in three different laboratories, then labora-tory could be either fixed or random, depending on whether the research question concerns differences between those specific laboratories or, more likely, that laboratory is an uninteresting source of non-independence.

Fixed and random effects are separately identified in the verbal equation specifying the model. For example:

$$Vigilance \sim PredationRisk + random(1|subject)$$

where the random effect is specified inside a pair of parentheses. The notation '1|subject' specifies that a **random intercept** is being fitted for each subject. This is analogous to centring each individual subject's data around its mean so that individual differences in mean vigilance are eliminated (illustrated graphically in Figure 11.4b).

A major advantage of the GLMM is that it provides a flexible framework within which it is possible to capture any structure in the data leading to non-independence. Complex structures can be accommodated by adding additional random effects. For instance, if the 20 birds comprised five groups of four siblings, two random effects could be specified in the model for family and subject, with subject nested within family.

More advanced extensions of this general approach are possible. For example, if the pedigree of a group of animals is known, it is possible to specify the relationships present in the pedigree in the random effects part of the model in order to control for non-independence due to different degrees of relatedness within a sample of animals (a so-called **animal model** [139]). If the data points in an analysis come from different species, the phylogenetic relationships between the species can be specified in the random effects part of the model in order to control for different degrees of evolutionary relatedness (a **phylogenetically controlled regression model** [140]).

11.8 Generalised Linear Models

As their name suggests, GLiMs are a generalisation of the GLM family. The GLiM relaxes the assumptions of the GLM that the outcome variable is continuously distributed and the residuals are normally distributed. By allowing for a variety of other distributions, GLiMs accommodate a wider range of outcome variables, including binary data, count data and proportions arising from counts [141].

Table 11.5 Examples of link functions and error distributions for modelling common data types with GLiM

Equivalent	Link function	Error distribution	Appropriate data types
GLM	Identity (i.e. straight line)	Normal	Continuous data with normally distributed residuals
Logistic regression	Logistic	Binomial	Categorical data (e.g. binary yes/no or 1/0 data)
Poisson regression	Log	Poisson	Count data (i.e. zeros and positive integers) with variance approximately equal to the mean

From a practical perspective, the GLiM eliminates the need to navigate the array of different statistical tests designed to cope with different data types. All can now be accommodated within one general framework. Models are specified in the same way as GLMs: $y \sim x_1 + x_2 + \ldots$, where the x variables can be categorical or continuous. The major difference from GLMs is the need to specify a **link function** and **error distribution** to account for the type of data being modelled. Common examples are given in Table 11.5.

Just as for GLMs, random effects can be added to GLiMs to create **generalised linear mixed-effects models** [142]. This type of model simultaneously accommodates non-normal outcome variables and suspected sources of non-independence by adding random effects to the GLiM.

11.8.1 What About Non-parametric Tests?

In the past, when the assumption of normality was violated, behavioural scientists turned to non-parametric statistical tests. The t-test was replaced with its non-parametric equivalent the Mann–Whitney U test, the paired t-test with the Wilcoxon test, the one-way ANOVA with the Kruskal–Wallis test, and so on [143]. Non-parametric tests were a mainstay in the analysis of behavioural data and they are still widely taught. However, the use of data transformations or GLiMs has largely eliminated the need for non-parametric tests, especially when sample sizes are fairly large. A major limitation of non-parametric tests is that they cannot cope with the complex data structures that are common in behavioural data and are readily accommodated by mixed-effects models.

11.9 Model Selection and Averaging

NHST examines whether the observed data supports or rejects the null hypothesis. An alternative approach is **model selection** – the idea that observed data can be used to adjudicate between several candidate models representing a range of hypotheses. NHST yields a binary conclusion whereby the null hypothesis is either rejected or retained. In contrast, model selection indicates the strength of support for each of the hypotheses compared. Model selection therefore aligns with the trend away from reporting findings in terms of the binary distinction between 'significant' and 'not significant'.

Model selection comes into its own when two or more predictor variables are highly correlated with one another ($r > 0.7$) and therefore cannot be included in the same model. Instead, separate models are compared, each containing one of the correlated predictors. Using NHST, several models may all lead to rejection of the null hypothesis, but this does not help in deciding which of the models is best. Model selection solves this problem by indicating which model is most strongly supported by the data.

Model selection is performed using the **Akaike information criterion (AIC)**, or related metrics, to compare the strength of support for each model in the set of models being compared [144]. AIC is a metric that is sensitive to how much of the variation in the data the predictors in the model explain, after accounting for how many predictors are included in the model. For a given number of predictors, the better the fit, the lower the AIC. For a given level of fit, the smaller the number of predictors, the lower the AIC. The model with the lowest AIC is the most strongly supported model. Simpler models with fewer predictors are favoured over more complex models because adding predictors to models generally improves model fit and can lead to **overfitting** of data.

Model selection produces a ranking of a set of models based on their AIC. Each model is additionally assigned a weight, based on its AIC, that can be interpreted as the percentage support for that model. For example, the top ranked model with the lowest AIC might be assigned a weight of 75 per cent, the second model 25 per cent and the third model 0 per cent. In the likely event that more than one model receives some support, it is not necessary to choose between these when the final results are presented. **Model averaging** uses the weight assigned to each model to compute weighted averages of the parameter estimates for the predictor variables in a set of models receiving non-zero support. Often, model averaging is restricted to the subset of best models that lie within some criterion distance (e.g. 2 AIC units) of the best model.

11.10 Multiverse Analysis

Data analysis involves a lot of decisions: which specific metric of the outcome variable to use, which data points to exclude, which transformations to use, which predictor variables to include in the statistical model, and so on. Some of these decisions will be informed by the research question. For example, when predictor variables x_1 and x_2 are correlated, the model $y \sim x_1$ addresses a different question from the model $y \sim x_1 + x_2$. Such decisions should be made at the planning stage and documented in the preregistration. In an ideal world, *all* necessary decisions would be made at the planning stage in order to reduce researcher degrees of freedom in the analysis and avoid p-hacking.

In practice, however, it is rarely possible to specify all decisions beforehand. For instance, it is impossible to know exactly which transformation will be necessary to correct normality and homogeneity of variance. Even when decisions have been preregistered, some may be arbitrary. For example, if data is aggregated prior to analysis, the specific statistic of central tendency used is an arbitrary choice. Exactly how many random effects are added to a model may also be somewhat arbitrary. Whenever alternative decisions are possible, especially if these are arbitrary, it is important to ask how sensitive the findings are to the decisions made. If a result is the same regardless of how the analysis was done, there is much greater confidence that it is a true finding.

Multiverse analysis assesses the robustness of results by estimating effects and p-values for an entire set of possible data analysis specifications. These methods include **vibration of effects** plots and **specification curves**, which provide visualisations of how effect sizes and p-values change according to the specification of the analysis [145].

11.11 Bayesian Methods

As noted earlier, the use of inferential statistics is largely about estimating the parameters of the population about which the researcher wishes to make inferences. When the only information available is the data in the sample measured, the best possible estimate of the mean of a variable in the population is the mean of the variable in the sample. The so-called **frequentist** modelling framework presented in this chapter is based on this fundamental premise. However, if *prior* information is available about the likely value of a parameter, **Bayesian inference** can be used to combine this prior information with information from the sample to derive optimum parameter estimates.

The Bayesian approach is attractive in theory. In practice, however, researchers often use non-specific assumptions about the prior information ('uninformed priors') in their models, with the result that there is no real gain over frequentist inference. Even so, one reason for preferring Bayesian methods is that they reverse the logic of NHST by allowing inferences to be made about the probability of a hypothesis given the data, rather than the other way around. This is an approach that many people find more intuitive [146]. However, frequentist approaches are currently easier to implement and more widely understood. Ultimately, it seems to be a matter of taste whether researchers prefer frequentist or Bayesian inference.

11.12 Confirmatory Versus Exploratory Analysis

Confirmatory analysis tests hypotheses set out in the study proposal or preregistration. **Exploratory analysis**, as the name suggests, allows for a more open-ended and creative exploration of the data collected. Exploratory analysis may or may not have been mentioned in the proposal or preregistration, but it is typically not associated with hypotheses or predictions made before collecting data.

Exploratory analysis is an important part of the research cycle because it often highlights results that form the basis for questions addressed in the next iteration of the cycle. The exploratory analysis of the current cycle becomes the confirmatory analysis of the next cycle, and so on.

Confirmatory and exploratory analyses are quite separate and distinct from descriptive and inferential statistics. Some authors equate exploratory analysis with descriptive statistics, but both descriptive and inferential statistics can be used in the service of both confirmatory and exploratory analyses.

11.13 Summary

- Statistical analysis is usually necessary to answer questions with behavioural data. Analysis should be planned and registered before collecting data.
- Once collected, a dataset should be formatted and permanently archived prior to analysis.
- Data is checked and visualised with descriptive statistics and graphs.
- Models representing hypotheses about the true effects present in the population from which the dataset is a sample are built and tested with inferential statistics.
- Many different hypotheses can be captured using a linear modelling framework in which an outcome variable is predicted with a combination of predictor variables and interactions.
- Sources of non-independence in datasets can be addressed with mixed models.
- The robustness of findings can be examined by comparing the results obtained when analysis is done in different ways using model selection and multiverse approaches.
- Confirmatory analysis designed to test preregistered hypotheses should be clearly differentiated from exploratory analysis that generates new hypotheses.

12

Interpreting and Communicating Findings

Science is a collaborative endeavour. Scientific progress depends on scientists being able to understand what has been done by other scientists, replicate their work and build on their findings. Communicating findings to the scientific community is therefore a vital part of doing science.

Scientists also often want to communicate their findings to the public. Behavioural research has implications for the public interest, in that it potentially bears on a wide range of matters including mental and physical health, disease transmission, security, the state of human society, climate change and the conservation of natural environments. Researchers have a social responsibility to form a realistic assessment of the potential implications of their work for society. They also have a responsibility to ensure the timely and appropriate communication of findings that are judged to be in the public interest.

Overselling results, negligence, bias and outright fraud have all contributed to the communication of findings that have not stood the test of time [15]. The publication and media reporting of findings that turn out to be not quite how they were presented or, at worst, not true, has damaging consequences for science and society. A prime example is the fraudulent study by Andrew Wakefield that suggested a link between the MMR (measles, mumps and rubella) vaccine and autism in children. Even though the original study published in 1988 was retracted, and numerous, much larger follow-up studies have found no support for Wakefield's claims, some parents are still reluctant to have their children vaccinated because of media coverage of the study. As a consequence, there has been a resurgence of measles and thousands of children are estimated to have been harmed.

Although some malpractice in science occurs at the stages of designing research and collecting data, much more occurs in the course of interpreting and communicating findings. The fraud in Wakefield's work did not emerge immediately, but the small sample size and poor study design should have raised red flags at the time of publication. Most cases of malpractice do not involve deliberate fraud or conscious attempts to misrepresent results;

rather, some scientists are simply unaware that certain common practices are wrong.

Our aims in this chapter are twofold. First, we outline some of the most common mistakes in the interpretation of data, and second, we provide guidance on how to communicate findings to other scientists and the public effectively and with integrity.

12.1 Common Statistical Mistakes

Researchers sometimes make claims that do not directly follow from their data. This often stems from incorrect use of statistics or incorrect interpretation of results. Here we list some of the most common mistakes in behavioural studies and suggest how to avoid them [147, 148].

One broad category is that of mistakes arising from misunderstandings and abuses of statistical significance testing:

- **Extraordinary findings from small studies are unlikely to be true.** Large effects obtained from small studies often fail to replicate because they are false positives (see section 2.3.2). Extraordinary findings based on small samples should be interpreted with caution and replicated before strong claims are made.
- **Non-significant findings in small studies are uninterpretable.** A non-significant result can mean one of two things: an underpowered study of a true effect (false negative), or a true null result (true negative). Without further information, it is impossible to tell which. A partial solution to this problem is to report effect sizes for non-significant effects. A non-significant small effect in a large study is unlikely to be important, whereas a non-significant moderate effect could warrant replication.
- **Statistical significance does not equal importance**. Statistical significance is based on an arbitrary criterion and depends on sample size (see section 11.5). The size of effects and their confidence intervals are more informative when judging the importance of a finding. Effect sizes and their confidence intervals should always be reported.
- **Relative measures of effect size can be misleading**. Relative rather than absolute measures of effect size can exaggerate effects. For example, a

50 per cent reduction in the relative risk of death following a drug treatment sounds huge – until it emerges that the absolute risk changed from, say, 2 in 10,000 in the control group to 1 in 10,000 in the drug group. It is more informative and less potentially misleading to report changes in absolute risk rather than relative risk.

- **Difference in significance does not imply significance of difference.** Just because a significant effect is found in one group and not in another when separate tests are conducted, does not mean the effect is significantly different in the two groups [149]. To establish whether an outcome variable is different between two groups, they must be directly compared with a single test (potentially followed by additional post-hoc analyses to discover exactly what is going on). More generally, it should be possible to test a hypothesis with a single statistical test. The use of multiple tests on different subsets of a dataset in the service of testing a single prediction should raise a red flag.

- **Multiple testing can produce false-positive results**. Analysing several slightly different versions of an outcome variable, calculating every possible correlation between a set of variables or performing every possible pairwise comparison within a set of variables are all examples of a dubious practice known as multiple testing or **fishing**. The danger of fishing for significant results is that if enough tests are conducted, some will be significant by chance alone. Multiple testing undermines the logic of NHST and inflates the probability of obtaining false-positive results. Multiple testing in confirmatory analyses should be eliminated by preregistering which outcome variable will be analysed, using the right statistical tests (e.g. a single ANOVA rather than multiple t-tests) and preregistering any planned comparisons. If multiple testing is undertaken as part of exploratory analysis, appropriate statistical methods should be used to account for the total number of tests conducted (e.g. the Bonferroni correction).

- **Pseudoreplication can produce false-positive results**. Failure to account for sources of non-independence in datasets, such as repeated measurements on the same individual, will tend to inflate the sample size, increase the precision of parameter estimates and increase the likelihood of false-positive results. Sources of non-independence must be accounted for in statistical analysis, either by taking means or by using appropriate statistical models (see section 11.7).

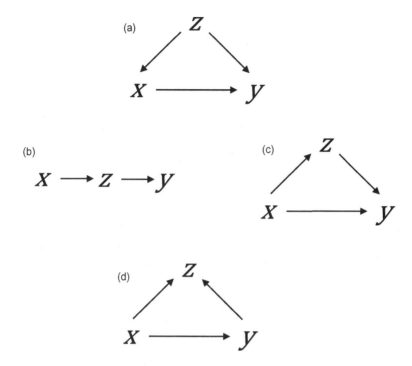

Figure 12.1 Alternative causal diagrams summarising how three variables, x, y and z, might causally affect one another. (a) Variable z is a confounder of the relationship between x and y. (b, c) Variable z is a mediator of the relationship between x and y. (d) Variable z is a collider in the relationship between x and y.

A second category of mistakes concerns inferences made from correlational data:

- **Correlations can be spurious.** In a small dataset, a significant correlation can result from a single outlying data point (see Figure 11.1d for an example). Data should always be plotted and analyses checked for the influence of outliers to avoid spurious correlations.
- **Correlation does not imply causation.** It is surprising how many researchers forget this basic statistical fact, especially when they have strong pre-existing reasons for believing in a causal link between two variables [150]. Alternative explanations for an association between two variables should always be considered (see section 4.4).
- **Bad statistical control produces errors in inference.** In observational studies where the aim is to determine the effect of predictor x on outcome

y, the addition of control variables to the statistical model can profoundly alter the results (see section 11.6). Controlling for a common cause of x and y (a **confounder**) reduces bias and improves inference about the true effect of x on y. In contrast, controlling for a variable that lies on the causal pathway between x and y (a **mediator** variable), or controlling for a common outcome of x and y (a **collider** variable), introduces bias in the estimation of the effect of x on y and can lead to false-negative and, more worryingly, false-positive errors (Figure 12.1) [151]. The lesson here is that more statistical control is not always better. Good statistical control is not straightforward and needs careful consideration. Drawing causal diagrams, such as those in Figure 12.1, is recommended when deciding which control variables to include in a regression model [152].

- **Differences in measurement error can produce errors in inference.** If a predictor variable x_1, which is the real cause of variation in outcome y, has been measured badly, and x_1 also affects another variable, x_2, which has been measured well, then regression might indicate that x_2 accounted for the variation in y better than x_1. This could lead to an incorrect conclusion that x_2 rather than x_1 was responsible for differences in y.

A third category of mistakes is that of **overgeneralisation**, when a researcher incorrectly extrapolates a result beyond what has actually been

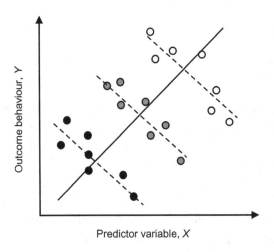

Figure 12.2 An illustration of Simpson's paradox. Within each group of points, there is a negative relationship between x and y, but across the whole dataset there is a positive relationship.

measured. It can be difficult deciding how far a given set of results can be generalised – that is, specifying the population to which the sample results refer. The temptation is often to overgeneralise – for example, by implying that the results of a laboratory experiment on learning in mice apply directly to learning in all animals, including humans. This general category of mistake appears in several forms:

- **Extrapolation beyond the range of values measured.** Just because a linear association is found between two variables in one part of their range does not mean this same association extends to all possible values.
- **Extrapolating from one context to another.** Just because an effect occurs in one context does not mean it will occur in a different context. For example, human cooperative behaviour measured in the laboratory using standard economic games may bear little relationship to how people behave in their natural environments.
- **Extrapolation from one level of analysis to another.** Different and sometimes opposite associations may exist at different levels of analysis. For example, within subjects there is often a trade-off between speed and accuracy of performance on behavioural tasks, leading to a negative association between speed and accuracy. Between subjects, however, speed and accuracy often have a positive association because more capable individuals tend to have higher absolute speed and accuracy than less capable individuals. Relationships of this type are known as **Simpson's paradox** (Figure 12.2).
- **Extrapolating from one species to another.** Animals are often used as models for understanding human behaviour. While this might sometimes make scientific sense, species clearly do differ, and attempts to generalise results from one species to another should be treated with caution. A light-hearted attempt to call out the rampant overgeneralisation of animal research in the media is the Twitter account @justsaysinmice which appends 'IN MICE' to hyped headline stories such as 'One-time treatment generates new neurons and eliminates Parkinson's disease'.

12.2 Writing a Scientific Paper

The scientific paper is still the most important means by which researchers communicate their findings. Scientific papers are the permanent record on

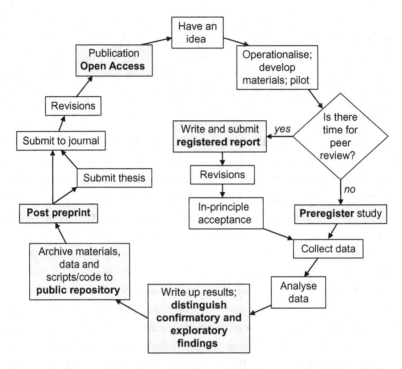

Figure 12.3 Steps in the research cycle showing where Open Science practices (shaded boxes) can be introduced to increase the openness and transparency of science and improve replicability.

which subsequent work builds. If a study is not published, then it might as well never have been done because its findings are not accessible to the scientific community. The adoption of Open Science practices has had a major impact on how scientific papers are written and published. Figure 12.3 summarises how Open Science practices are integrated into the research cycle.

Scientific papers are aimed primarily at other researchers in the same field, but changes in publishing practices are expanding their potential audience. The widespread availability of papers online, the increase in non-specialist journals, and the use of search engines and social media to discover relevant work all mean that scientists are now more likely to encounter papers outside their field. As a result of preprints and **Open Access publishing**, non-scientists, including journalists, policy makers and the public, have free access to a growing proportion of primary empirical research. These developments make it even more important for scientific papers to be written in clear, unambiguous and accessible language.

One easily avoidable barrier to effective communication is a stilted writing style. For example, it is better not to use the impersonal passive voice in formal scientific writing. The impersonal passive voice (as in 'It is argued...' or 'It is believed...') is ambiguous because it is not clear *who* is arguing or believing. If it is the authors, then it is clearer to write: 'We argue...' or 'I believe...'. If it is someone else, the source should be cited. The *Style Guide for Authors and Editors* of the British Psychological Society provides succinct advice on good science writing [153].

Plagiarism – directly copying text from previously published sources without attribution – takes many forms, including republishing someone else's paper and copying or paraphrasing substantial parts of another paper or indeed one's own earlier paper (self-plagiarism). Plagiarism of any kind is unethical. Self-plagiarism may seem harmless, but it can infringe copyright. Most journals require authors to confirm that the submitted work has not been published previously and use plagiarism-checking software.

There is no universal formula for writing a scientific paper. However, most papers will include the following elements.

12.2.1 Title and Abstract

The title should be succinct but descriptive. The abstract, typically of 150–250 words, is important because, along with the title, it is likely to be the first, and often the *only*, part of the paper that is actually read. Many meta-analysis protocols involve scanning titles and abstracts to determine whether a published paper is a candidate for inclusion. It is therefore critical that the title and abstract do not misrepresent the contents and provide as much relevant detail as possible, including the species studied. Ideally, results should be reported with estimates of effect size.

Graphical abstracts are a recent innovation in some journals, where authors are additionally invited to prepare a single image that captures the main finding of the paper. Graphical abstracts can be useful for dissemination on social media.

12.2.2 Introduction

The introduction should provide the background information necessary to understand the study. It can be helpful to divide the introduction into paragraphs answering the following questions: (1) What is the general question addressed by the study? (2) What is the current knowledge relating to this question? (3) What is the gap in current knowledge that the study is designed to fill? (4) How will the study achieve the stated aim? The final paragraph typically ends with a clear statement of the hypotheses and predictions. If the study has been preregistered, then this should be mentioned. Confirmatory and exploratory hypotheses should be clearly identified in order to avoid suspicions of HARKing (Hypothesising After Results are Known).

12.2.3 Methods

The methods should be described in sufficient detail to allow another researcher to replicate the study. Inadequate reporting of methods is recognised as a major barrier to replication.

Ethical approvals should be described, including the granting bodies and reference numbers. Details of the location of any preregistration of the study should be given. Studies involving captive animals should provide details of animal housing and husbandry conditions and any steps taken in scientific procedures to reduce pain, suffering and distress.

The **ARRIVE** (Animal Research: Reporting of In Vivo Experiments) guidelines (arriveguidelines.org) are a checklist of the information that should be reported in any scientific paper documenting research on live animals [154]. The checklist includes the minimum information required to enable proper scrutiny of the quality of research and replication by other scientists. Although the ARRIVE guidelines refer specifically to experimental research on animals, they have wider applicability to observational studies and studies on humans. The guidelines have been endorsed by more than 1000 journals. Table 12.1 reproduces the 'Essential 10' checklist from the ARRIVE guidelines [154].

If the study involves coding of behaviour, full descriptions of behavioural categories and how they were recorded should be included. Any non-

Table 12.1 The ARRIVE guidelines 'Essential 10' checklist (reproduced and modified from [154])

1	Study design	For each experiment, provide brief details of study design including: (a) The groups being compared, including control groups. If no control group has been used, the rationale should be stated. (b) The experimental unit (e.g. a single animal, litter or cage of animals).
2	Sample size	(a) Specify the exact number of experimental units allocated to each group, and the total number in each experiment. Also indicate the total number of animals used. (b) Explain how the sample size was decided. Provide details of any a priori sample size calculation, if done.
3	Inclusion and exclusion criteria	(a) Describe any criteria used for including and excluding animals (or experimental units) during the experiment, and data points during the analysis. Specify whether these criteria were established a priori. If no criteria were set, state this explicitly. (b) For each experimental group, report any animals, experimental units or data points not included in the analysis and explain why. If there were no exclusions, state so. (c) For each analysis, report the exact sample size in each experimental group.
4	Randomisation	(a) State whether randomisation was used to allocate experimental units to control and treatment groups. If done, provide the method used to generate the randomisation sequence. (b) Describe the strategy used to minimise potential confounders such as the order of treatments and measurements, or animal/cage location. If confounders were not controlled, state this explicitly.
5	Blinding	Describe who was aware of the group allocation at the different stages of the experiment (during the allocation, the conduct of the experiment, the outcome assessment and the data analysis).
6	Outcome measures	(a) Clearly define all outcome measures assessed (e.g. cell death, molecular markers or behavioural changes). (b) For hypothesis-testing studies, specify the primary outcome measure, i.e. the outcome measure that was used to determine the sample size.

Table 12.1 (*cont.*)

7	Statistical methods	(a) Provide details of the statistical methods used for each analysis, including software used. (b) Describe any methods used to assess whether the data met the assumptions of the statistical approach, and what was done if the assumptions were not met.
8	Experimental animals	(a) Provide species-appropriate details of the animals used, including species, strain and substrain, sex, age or developmental stage and, if relevant, weight. (b) Provide further relevant information on the provenance of animals, health/immune status, genetic modification status, genotype and any previous procedures.
9	Experimental procedures	For each experimental group, including controls, describe the procedures in enough detail to allow others to replicate them, including: (a) What was done, how it was done and what was used. (b) When and how often. (c) Where (including detail of any acclimatisation periods). (d) Why (provide rationale for procedures).
10	Results	For each experiment conducted, including independent replications, report: (a) Summary/descriptive statistics for each experimental group, with a measure of variability where applicable (e.g. mean and standard deviation, or median and range). (b) If applicable, the effect size with a confidence interval.

standard stimuli or other materials, such as novel questionnaires, should either be reproduced in the paper or made available in full as supplementary information published alongside the paper.

12.2.4 Results

Empirical findings should be presented in this section without interpretation. It is essential to distinguish clearly between confirmatory and exploratory analysis (see section 11.12).

The **SAMPL** (Statistical Analyses and Methods in the Published Literature) guidelines provide advice on the correct reporting of statistical analyses [155]. These guidelines are based on two principles: first, that statistical methods are described in enough detail to enable a knowledgeable reader with access to the original data to verify the reported results, and second, that sufficient detail is provided that the results can be incorporated into other analyses, such as meta-analyses. For every model fitted, the outcome variable should be stated, along with all the predictor variables included in the model. The exact specification of the test or model should be given, including any interaction terms and random effects (this can be done using a verbal equation; see section 11.5.1). The sample size for each analysis should be clearly stated. Effect sizes (or parameter estimates), their confidence intervals and the outcome of any significance testing should be reported for all predictor variables in every test or model.

Publishing the full dataset and a statistical analysis script (e.g. an R script) alongside a paper is an effective way of providing a transparent analysis pipeline and is increasingly regarded as best practice. The data and script should be made available to referees when the paper is first submitted to a journal to facilitate checking.

Visual presentation of quantitative data is extremely important. A graph or other form of visual display is generally much more informative than densely packed tables of numbers. Many different types of visual display are available via packages such as R, and it is worth spending time exploring the best way to present a result [156]. We suggest the following general guidelines for plotting data.

In graphs, the outcome variable is conventionally drawn on the y-axis, with the predictor variable on the x-axis. Each axis should be clearly labelled with the name of the variable (e.g. average distance travelled, vocalisation rate, amount of food consumed) and the units in which the variable is measured (e.g. km d^{-1}, min^{-1}, g h^{-1}). Where a metric is expressed as the total number of occurrences, the period of time over which the events were counted should be indicated (e.g. per 30 min). The type of metric used for behavioural categories should be clearly indicated, using reciprocal units of time (e.g. s^{-1}, min^{-1}, d^{-1}) for frequency metrics and units of time (e.g. s, min, d) for duration and latency metrics. Time sampling measures and the proportion of time spent performing an activity are expressed as dimensionless scores between 0 and 1.

Raw data should be plotted wherever possible. The Gardner–Altman plot shown in Figure 11.2a displays a huge amount of information compared with a conventional bar graph. Box plots and their variants are more informative than bar graphs for large datasets because they provide more information about the distribution of the data.

When average (as opposed to raw) values are plotted, as in a conventional bar graph, it is helpful to indicate the variation about the average with error bars. A common practice is to plot an error bar proportional in length to one standard error above and below the mean. Variability about a median is often indicated with bars denoting the IQR (which can be asymmetrical).

Results are often presented in the form of multiple plots, all sharing the same x-axis but each with a different y-axis. This type of plot can be misleading when casually inspected, unless attention is drawn to the fact that the y-axis scales differ. It is doubly important in such cases that the axes are clearly labelled.

Graphs with a restricted range for the y-axis accentuate small variations in the outcome variable y and can be misleading. It may be justifiable to focus on a particular part of the y-axis range to see patterns in the data, but it should be made clear that the y-axis shows only a selected portion of the range.

Graphs should not be cluttered with distracting backgrounds, three-dimensional effects or shadows that add no useful information. Colour should be used sparingly and only to enhance interpretation; colour palettes are available that are visible to colour-blind readers.

12.2.5 Discussion

The main purpose of this section is to interpret the results with reference to the study's stated aims and hypotheses, relate the results to other relevant studies in the literature, and comment on their theoretical and practical implications. The discussion should be concise and not excessively speculative.

It may be helpful to divide the discussion into paragraphs that address the same four questions used to structure the introduction, in reverse order. The first paragraph briefly reiterates what was done and summarises the main findings. The second paragraph discusses whether and how the results have filled the knowledge gap. The third paragraph discusses any other secondary results, such as replication of established effects and how they relate to the existing literature. The final paragraph considers the implications for science and society and summarises the take-home message of the paper.

The discussion should comment on the study's limitations and highlight areas where further work is required. Some journals require an explicit 'limitations' paragraph. Any uncertainty about the findings should be communicated honestly.

12.2.6 References

All literature cited in a paper must be properly itemised in the reference list. Annoyingly, scientific journals do not use a standard style for citations and reference lists. Reference management software makes it much easier to reformat references for different journals.

12.3 Publishing a Scientific Paper

Writing a paper is only the first step in getting work published in a peer-reviewed academic journal. The publication process requires a number of decisions.

12.3.1 Authorship

It will be necessary to decide who will be the named authors on a paper and in what order their names will appear. Various organisations provide advice on the criteria for authorship, and there is reasonable agreement on what is necessary. For example, the International Committee of Medical Journal Editors suggests the following criteria for authorship [157]:

- Substantial contributions to the conception or design of the work; or the acquisition, analysis or interpretation of data for the work; AND
- Drafting the work or revising it critically for important intellectual content; AND

- Final approval of the version to be published; AND
- Agreement to be accountable for all aspects of the work in ensuring that questions related to the accuracy or integrity of any part of the work are appropriately investigated and resolved.

Scientists who have contributed to a study but who do not meet all of the above requirements should be listed in the acknowledgements. To improve transparency and discourage abuses, many journals now have an 'author contributions' statement in which the role of every author is explicitly described.

The order of authors has implications for how credit is attributed. On a multi-author paper, first authorship typically indicates the person who has done most of the work, while the last author is typically the senior scientist who takes overall responsibility for the study. One author will be designated as the 'corresponding author' who will act as the point of contact and answer questions about the paper.

12.3.2 Preprinting

The publication process may take many months, sometimes years. It may be necessary to go through multiple rounds of peer review and multiple journals before a paper is finally published. These delays can seem interminable and they can have a serious impact on careers and the speed of scientific progress. **Preprints** offer a solution by making research publicly available as soon as a paper is written. Preprints are non-peer-reviewed versions of papers that have their own doi (digital object identifier) and can be found online and cited. All reputable journals now accept work that has appeared previously in preprint form, which means that preprinting does not prevent publication in a peer-reviewed journal. Several free preprint servers accept various types of behavioural research (Table 12.2).

12.3.3 Choosing a Scientific Journal

Hundreds of different journals publish behavioural research. Choosing the right one will depend on a number of factors, including the subject area, the quality and importance of the science, and the authors' priorities.

Table 12.2 Major preprint servers covering behavioural sciences

Name	URL	Coverage
PsyArXiv	psyarxiv.com	Psychology
bioRXiv	biorxiv.org	Bioscience, including: animal behaviour, cognition, neuroscience
medRXiv	medrxiv.org	Health sciences, including: addiction, clinical psychology, epidemiology, health policy, nutrition, psychiatry
SocArXiv	socopen.org	Anthropology, economics, political science, sociology

Whether the published paper will be Open Access (i.e. free for anyone to read for perpetuity) is an increasingly important consideration. Many funders now require Open Access publication, and scientists benefit from making their work easier to access. On average, Open Access papers attract more citations. The downside of Open Access publication is that someone has to pay – typically the author or their institution – although some Open Access journals offer waivers to certain categories of scientists.

The selectivity of a journal is also a primary consideration. The most prestigious journals are highly selective and reject most submissions without review. Importantly, this selectivity is not based solely on the quality of the science; it also reflects the editor's opinion of the work's likely importance and impact. Highly selective journals are consequently less likely to publish negative results and replications, no matter how well the science has been done. In contrast, some of the newer Open Access journals have explicit policies of only assessing papers on the quality of the science and whether the research has been done ethically. Top journals often invite **pre-submission enquiries**. These can be a useful way of gauging an editor's likely interest before spending time on formally submitting the full paper.

12.3.4 Responding to Reviewers' Comments

A journal editor will send a submitted paper to two or more scientists for peer review. Based on their comments, the editor will accept the paper for publication (usually subject to minor or major revisions), invite it for resubmission (subject to revisions) or reject it outright.

Reviewers often make suggestions that substantially improve a paper. However, they sometimes make criticisms arising from misunderstanding. This can be frustrating for the authors, but the fact that a referee has misunderstood something is important information. The paper will almost certainly be improved by identifying the source of the misunderstanding and explaining the science more clearly. Misunderstandings are particularly common in inter-disciplinary behavioural research because terms can have different meanings in different fields. A 'functional explanation' means something different to behavioural ecologists and cognitive neuroscientists, while 'adaptation' refers to different phenomena in evolutionary biology and physiology.

12.4 Other Forms of Scientific Communication

Publishing papers is not the only way of communicating scientific findings. Conference presentations provide an opportunity to present work to the scientific community and obtain feedback from colleagues. The networking that takes place at conferences is often crucial for establishing new collaborations.

Conference presentations take the form of posters or talks. A poster or talk should not be approached as if it were a direct oral presentation of a scientific paper; they are different forms of communication. The key to doing them well is to present the minimum detail necessary to make the main findings intelligible.

12.5 Communicating with the Public

Scientists should be alert to the wider societal implications of their work, both positive and negative. Results with significant implications for the public interest are likely to attract media attention, whether or not this is sought, and researchers should be prepared for this possibility.

Many organisations, including journals, ethical review committees and grant-giving bodies, require a **non-technical (lay) summary** of research. This is a summary designed for the public, written in language that can be understood by intelligent non-scientists and avoiding jargon. In practice, non-technical summaries often have an important function in explaining research to scientists from other fields.

If a study seems likely to attract media attention, it is wise to prepare a **press release** before publication. The aim of a press release is to influence how journalists portray a study. A press release differs from a non-technical summary in additionally attempting to pre-empt some of the questions that journalists are likely to ask with considered responses. A carefully worded press release is particularly important if the findings are controversial or have the potential to be misunderstood or abused. Most journalists will draw on the material in a press release if one is available. Scientists therefore have a responsibility to ensure that the information in a press release is accurate and not hyped or overgeneralised.

Writing good non-technical summaries and press releases is a skill. Many institutions have media officers who can assist with dealing with the media and writing and filing press releases. Funding bodies and universities increasingly provide training courses to help scientists develop competence in public communication. Advice can also be obtained from organisations such as the Science Media Centre (sciencemediacentre.org). Training in public engagement is widely regarded as a core element of continuing professional development for scientists.

12.6 Competing Interests

Competing interests, or conflicts of interests (COIs), are a ubiquitous problem in science. In the context of research, a COI can be defined as any situation in which a scientist has multiple interests – financial, political or otherwise – one or more of which has the potential to distort their motivation to conduct, interpret and communicate their research objectively.

The presence of a COI is not in itself evidence of any improper behaviour, but it does provide a context that can be important in interpreting the work. For example, any research showing that smoking tobacco is less harmful is likely to be treated with greater scepticism if it was funded by the tobacco industry, or if the lead author is known to be a heavy smoker (as was the case for the statistician Ronald Fisher who disputed the evidence that smoking causes cancer).

Journals typically require authors to declare any potential COIs in a paper. These are usually restricted to specific sources of funding, such as companies or charities that could be perceived to have a particular agenda in relation to the research. The fact that the majority of scientists rely on publications for getting jobs, promotions and research grants is now widely acknowledged as a ubiquitous COI in science, albeit one that is rarely explicitly declared. The pressure to publish papers, and ideally attention-grabbing papers in top journals, undoubtedly reinforces the replication crisis. Workable solutions to this problem are yet to emerge.

12.7 Integrity and Responsibility

Honesty in research is crucial, both for the sake of scientists' individual reputations and for the reputation of science as a whole. In the USA, the Office for Research Integrity (ori.hhs.gov) oversees the conduct of publicly funded biomedical and behavioural research. Among other things, it promotes good practice, issues guidelines, monitors compliance and investigates cases of misconduct. Other countries have comparable bodies, such as the UK Research Integrity Office (ukrio.org). The Committee on Publication Ethics (COPE: publicationethics.org) promotes integrity in scholarly research and publications. It provides advice, aimed primarily at journal editors but useful also for authors, on ethical issues such as plagiarism, falsification, authorship, COIs and misconduct.

When cases of research misconduct or fraud are revealed, public trust in science is eroded. Every scientist needs to recognise how much the whole enterprise depends on their own behaviour. It is more important than ever for each individual scientist to stick firmly to the classical scientific virtues of honesty, scepticism and integrity at every stage of research. The replication crisis has reinforced the requirement for transparency and openness in the way science is done.

12.8 Summary

- Interpreting results correctly and communicating them honestly are vital parts of what scientists do.

- Incorrect interpretation of data often results from avoidable statistical mistakes. Common pitfalls arise from abuse of significance testing, misunderstanding of correlations and overgeneralisation of findings.
- Publishing peer-reviewed papers in scientific journals is the primary means by which researchers communicate their findings to other scientists. A scientific paper has an established basic format comprising title, abstract, introduction, methods, results and discussion.
- Open Science practices are an important part of the modern publication process.
- Non-technical (lay) summaries and press releases are tools for communicating behavioural research to journalists and the public.
- All science involves potential conflicts of interest, and their influence on scientific communication is an unresolved cause for concern.
- Several organisations oversee the integrity of science, but ultimately it is the personal responsibility of each individual researcher to behave with openness and integrity.

References

1. A. Trewavas, Intelligence, cognition, and language of green plants. *Frontiers in Psychology*, **7** (2016), 1–9.

2. I. Rahwan, M. Cebrian, N. Obradovich, *et al.*, Machine behaviour. *Nature*, **568** (2019), 477–486.

3. R. F. Baumeister, K. D. Vohs and D. C. Funder, Psychology as the science of self-reports and finger movements: whatever happened to actual behavior? *Perspectives on Psychological Science*, **2** (2007), 396–403.

4. D. A. Schoeller, Limitations in the assessment of dietary energy intake by self-report. *Metabolism*, **44** (1995), 18–22.

5. R. L. Monk, D. Heim, A. Qureshi and A. Price, "I have no clue what I drunk last night". Using smartphone technology to compare in-vivo and retrospective self-reports of alcohol consumption. *PLOS ONE*, **10** (2015) e0126209.

6. J. Williams, C. Stönner, J. Wicker, *et al.*, Cinema audiences reproducibly vary the chemical composition of air during films, by broadcasting scene specific emissions on breath. *Scientific Reports*, **6** (2016), 1–10.

7. J. M. Clarkson, D. M. Dwyer, P. A. Flecknell, *et al.*, Handling method alters the hedonic value of reward in laboratory mice. *Scientific Reports*, **8** (2018), 2448.

8. R. E. Sorge, L. J. Martin, K. A. Isbester, *et al.*, Olfactory exposure to males, including men, causes stress and related analgesia in rodents. *Nature Methods*, **11** (2014), 629–632.

9. D. Nettle, M. Joly, E. Broadbent, *et al.*, Opportunistic food consumption in relation to childhood and adult food insecurity: an exploratory correlational study. *Appetite*, **132** (2018), 222–229.

10. H. Pashler and E. J. Wagenmakers, Editors' introduction to the special section on replicability in psychological science: a crisis of confidence? *Perspectives on Psychological Science*, **7** (2012), 528–530.

11. M. Baker and D. Penny, Is there a reproducibility crisis? *Nature*, **533** (2016), 452–454.

12. Open Science Collaboration, Estimating the reproducibility of psychological science. *Science*, **349** (2015), acc4716.

13. C. F. Camerer, A. Dreber, F. Holzmeister, *et al.*, Evaluating the replicability of social science experiments in *Nature* and *Science* between 2010 and 2015. *Nature Human Behaviour*, **2** (2018), 637–644.

14. R. A. Klein, M. Vianello, F. Hasselman, *et al.*, Many Labs 2: investigating variation in replicability across samples and settings. *Advances in Methods and Practices in Psychological Science*, **1** (2018), 443–490.

15. S. Ritchie, *Science Fictions: Exposing Fraud, Bias, Negligence and Hype in Science.* (London: Penguin, 2020).

16. D. E. Kroodsma, Suggested experimental designs for song playbacks. *Animal Behaviour*, **37** (1989), 600–609.

17. M. R. Munafò and G. D. Smith, Repeating experiments is not enough. *Nature*, **553** (2018), 399–401.

18. C. G. Begley and J. P. A. Ioannidis, Reproducibility in science: improving the standard for basic and preclinical research. *Circulation Research*, **116** (2015), 116–126.

19. J. P. A. Ioannidis, Why most published research findings are false. *PLOS Medicine*, **2** (2005), e124.

20. C. Bodden, V. T. von Kortzfleisch, F. Karwinkel, *et al.*, Heterogenising study samples across testing time improves reproducibility of behavioural data. *Scientific Reports*, **9** (2019), 8247.

21. B. Voelkl and H. Würbel, Reproducibility crisis: are we ignoring reaction norms? *Trends in Pharmacological Sciences*, **37** (2016), 509–510.

22. K. S. Button, J. P. A. Ioannidis, C. Mokrysz, *et al.*, Power failure: why small sample size undermines the reliability of neuroscience. *Nature Reviews Neuroscience*, **14** (2013), 365–376.

23. D. J. Benjamin, J. O. Berger, M. Johannesson, *et al.*, Redefine statistical significance. *Nature Human Behaviour*, **2** (2018), 6–10.

24. J. P. Simmons, L. D. Nelson and U. Simonsohn, False-positive psychology: undisclosed flexibility in data collection and analysis allows presenting anything as significant. *Psychological Science*, **22** (2011), 1359–1366.

25. D. Fanelli, How many scientists fabricate and falsify research? A systematic review and meta-analysis of survey data. *PLOS ONE*, **4** (2009), e5738.

26. M. R. Munafò, B. A. Nosek, D. V. M. Bishop, *et al.*, A manifesto for reproducible science. *Nature Human Behaviour*, **1** (2017), 0021.

27. M. Borenstein, L. V. Hedges, J. P. Higgins and H. R. Rothstein, *Introduction to Meta-Analysis.* (Chichester, UK: John Wiley & Sons, 2009).

28. F. C. Fang and A. Casadevall, Retracted science and the retraction index. *Infection and Immunity*, **79** (2011), 3855–3859.

29. F. C. Fang, R. G. Steen and A. Casadevall, Misconduct accounts for the majority of retracted scientific publications. *Proceedings of the National Academy of Sciences USA*, **109** (2012), 17028–17033.

30. B. K. Redman, H. N. Yarandi and J. F. Merz, Empirical developments in retraction. *Journal of Medical Ethics*, **34** (2008), 807–809.

31. S. Crüwell, J. van Doorn, A. Etz, *et al.*, Seven easy steps to open science: an annotated reading list. *Zeitschrift für Psychologie*, **227** (2019), 237–248.

32. W. E. Frankenhuis and D. Nettle, Open Science is liberating and can foster creativity. *Perspectives on Psychological Science*, **13** (2018), 439–447.

33. B. A. Nosek, G. Alter, G. C. Banks, *et al.*, Promoting an open research culture. *Science*, **348** (2015), 1422–1425.

34. N. Tinbergen, On aims and methods of ethology. *Zeitschrift für Tierpsychologie*, **20** (1963), 410–433.

35. P. Bateson and K. N. Laland, Tinbergen's four questions: an appreciation and an update. *Trends in Ecology and Evolution*, **28** (2013), 712–718.

36. R. M. Nesse, Tinbergen's four questions, organized: a response to Bateson and Laland. *Trends in Ecology and Evolution*, **28** (2013), 681–682.

37. E. Jonas and K. P. Kording, Could a neuroscientist understand a microprocessor? *PLOS Computational Biology*, **13** (2017), e1005268.

38. J. Henrich, S. J. Heine and A. Norenzayan, The weirdest people in the world? *Behavioral and Brain Sciences*, **33** (2010), 61–83.

39. D. Nettle, C. Andrews and M. Bateson, Food insecurity as a driver of obesity in humans: the insurance hypothesis. *Behavioral and Brain Sciences*, **40** (2017), e105.

40. A. Dubiec, I. Góźdź and T. D. Mazgajski, Green plant material in avian nests. *Avian Biology Research*, **6** (2013), 133–146.

41. S. T. Bate and R. A. Clark, *The Design and Statistical Analysis of Animal Experiments*. (Cambridge, UK: Cambridge University Press, 2014).

42. S. E. Lazic, *Experimental Design for Laboratory Biologists: Maximising Information and Improving Reproducibility*. (Cambridge, UK: Cambridge University Press, 2016).

43. G. D. Ruxton and N. Colegrave, *Experimental Design for the Life Sciences*, 2nd edn. (Oxford, UK: Oxford University Press, 2006).

44. M. Bateson and D. Nettle, Why are there associations between telomere length and behaviour? *Philosophical Transactions of the Royal Society of London B: Biological Sciences*, **373** (2018), 20160438.

45. W. W. Cruze, Maturation and learning in chicks. *Journal of Comparative Psychology*, **19** (1935), 371–408.

46. N. Tinbergen, G. J. Broekhuysen, F. Feekes, *et al.*, Egg shell removal by the black-headed gull, *Larus ridibundus* L.; a behaviour component of camouflage. *Behaviour*, **19** (1962), 74–116.

47. N. Percie du Sert, I. Bamsey, S. T. Bate, *et al.*, The Experimental Design Assistant. *PLOS Biology*, **15** (2017), e2003779.

48. M. Rigdon, K. Ishii, M. Watabe and S. Kitayama, Minimal social cues in the dictator game. *Journal of Economic Psychology*, **30** (2009), 358–367.

49. R. Shaw, M. F. W. Festing, I. Peers and L. Furlong, Use of factorial designs to optimize animal experiments and reduce animal use. *ILAR Journal*, **43** (2002), 223–232.

50. S. R. Brand, S. M. Engel, R. L. Canfield and R. Yehuda, The effect of maternal PTSD following in utero trauma exposure on behavior and temperament in the 9-month-old infant. *Annals of the New York Academy of Sciences*, **1071** (2006), 454–458.

51. G. D. Smith, Mendelian randomization for strengthening causal inference in observational studies: application to gene × environment interactions. *Perspectives on Psychological Science*, **5** (2010), 527–545.

52. L. Rode, S. E. Bojesen, M. Weischer and B. G. Nordestgaard, High tobacco consumption is causally associated with increased all-cause mortality in a general population sample of 55 568 individuals, but not with short telomeres: a Mendelian randomization study. *International Journal of Epidemiology*, **43** (2014), 1473–1483.

53. J. Cohen, A power primer. *Psychological Bulletin*, **112** (1992), 155–159.

54. F. Faul, E. Erdfelder, A. G. Lang and A. Buchner, G*Power 3: a flexible statistical power analysis program for the social, behavioral, and biomedical sciences. *Behavior Research Methods*, **39** (2007), 175–191.

55. E. Erdfelder, F. Faul, A. Buchner and A. G. Lang, Statistical power analyses using G*Power 3.1: tests for correlation and regression analyses. *Behavior Research Methods*, **41** (2009), 1149–1160.

56. K. Neumann, U. Grittner, S. K. Piper, *et al.*, Increasing efficiency of preclinical research by group sequential designs. *PLOS Biology*, **15** (2017), e2001307.

57. J. P. Garner, Stereotypies and other abnormal repetitive behaviors: potential impact on validity, reliability, and replicability of scientific outcomes. *ILAR Journal*, **46** (2005), 106–117.

58. E. Drinkwater, E. J. H. Robinson and A. G. Hart, Keeping invertebrate research ethical in a landscape of shifting public opinion. *Methods in Ecology and Evolution*, **10** (2019), 1265–1273.

59. Association for the Study of Animal Behaviour/Animal Behaviour Society. Guidelines for the treatment of animals in behavioural research and teaching. *Animal Behaviour*, **135** (2018), I–X.

60. The British Psychological Society, *Guidelines for Psychologists Working with Animals*. (Leicester, UK: The British Psychological Society, 2012).

61. A. Brønstad, C. Newcomer, T. Decelle, *et al.*, Current concepts of harm–benefit analysis of animal experiments – Report from the AALAS-FELASA working group on harm–benefit analysis – Part 1. *Laboratory Animals*, **50** (Suppl.) (2016), 1–20.

62. American Psychological Association. Ethical Principles of Psychologists and Code of Conduct. *American Psychologist*, **57** (2017), 1060–1073.

63. P. Bateson, Ethics and behavioral biology. *Advances in the Study of Behavior*, **35** (2005), 211–233.

64. J. W. Driscoll and P. Bateson, Animals in behavioural research. *Animal Behaviour*, **36** (1988), 1569–1574.

65. I. Cuthill, Field experiments in animal behaviour: methods and ethics. *Animal Behaviour*, **42** (1991), 1007–1014.

66. W. M. S. Russell and R. L. Burch, *The Principles of Humane Experimental Technique*. (London: Methuen, 1959).

67. R. Hubrecht and J. Kirkwood, *The UFAW Handbook on the Care and Management of Laboratory and Other Research Animals*, 8th edn. (Chichester, UK: John Wiley & Sons, 2010).

68. A. J. Smith, R. E. Clutton, E. Lilley, *et al.*, PREPARE: guidelines for planning animal research and testing. *Laboratory Animals*, **52** (2018), 135–141.

69. A. J. Smith, R. E. Clutton, E. Lilley, *et al.*, *The PREPARE Guidelines Checklist*. (Oslo, Norecopa, 2020). norecopa.no/media/7893/prepare_checklist_english.pdf

70. The British Psychological Society, *Code of Human Research Ethics*. (Leicester, UK: British Psychological Society, 2014).

71. The British Psychological Society, *Ethics Guidelines for Internet-Mediated Research*. (Leicester, UK: British Psychological Society, 2017).

72. M. C. Rousu, G. Colson, J. R. Corrigan, *et al.*, Deception in experiments: towards guidelines on use in applied economics research. *Applied Economic Perspectives and Policy*, **37** (2015), 524–536.

73. D. J. Cooper, A note on deception in economic experiments. *Journal of Wine Economics*, **9** (2014), 111–114.

74. European Parliament and Council of European Union, *Regulation (EU) 2016/679*, 27 April 2016.

75. C. M. O'Keefe and D. B. Rubin, Individual privacy versus public good: protecting confidentiality in health research. *Statistics in Medicine*, **34** (2015), 3081–3103.

76. H. Gray, H. Bertrand, C. Mindus, *et al.*, Physiological, behavioral, and scientific impact of different fluid control protocols in the rhesus macaque (*Macaca mulatta*). *eNeuro*, **3** (2016), ENEURO.0195-16.2016.

77. G. Schino, G. Perretta, A. M. Taglioni, *et al.*, Primate displacement activities as an ethopharmacological model of anxiety. *Anxiety*, **2** (1996), 186–191.

78. C. Poirier, C. J. Oliver, P. Flecknell and M. Bateson, Pacing behaviour in laboratory macaques is an unreliable indicator of acute stress. *Scientific Reports*, **9** (2019), 7476.

79. S. S. Stevens, On the theory of scales of measurement. *Science*, **103** (1946), 677–680.

80. C. Hewson, Conducting research on the internet – a new era. *The Psychologist*, **27** (2014), 946–950.

81. D. Wahlsten, N. R. Rustay, P. Metten and J. C. Crabbe, In search of a better mouse test. *Trends in Neurosciences*, **26** (2003), 132–136.

82. A. Ennaceur and P. L. Chazot, Preclinical animal anxiety research – flaws and prejudices. *Pharmacology Research & Perspectives*, **4** (2016), e00223.

83. M. V. Hernández-Lloreda and F. Colmenares, The utility of generalizability theory in the study of animal behaviour. *Animal Behaviour*, **71** (2006), 983–988.

84. D. P. Mersch, A. Crespi and L. Keller, Tracking individuals shows spatial fidelity is a key regulator of ant social organization. *Science*, **340** (2013), 1090–1093.

85. S. Ransdell and F. Lauderdale, Teaching psychology as a laboratory science in the age of the internet. *Behavior Research Methods, Instruments, & Computers*, **34** (2002), 145–150.

86. G. Stoet, PsyToolkit: a software package for programming psychological experiments using Linux. *Behavior Research Methods*, **42** (2010), 1096–1104.

87. G. Stoet, PsyToolkit: a novel web-based method for running online questionnaires and reaction-time experiments. *Teaching of Psychology*, **44** (2017), 24–31.

88. O. Friard and M. Gamba, BORIS: a free, versatile open-source event-logging software for video/audio coding and live observations. *Methods in Ecology and Evolution*, **7** (2016), 1325–1330.

89. L. Hänninen and M. Pastell, CowLog: open-source software for coding behaviors from digital video. *Behavior Research Methods*, **41** (2009), 472–476.

90. R. P. Wilson, M. D. Holton, A. di Virgilio, *et al.*, Give the machine a hand: a Boolean time-based decision-tree template for rapidly finding animal behaviours in multisensor data. *Methods in Ecology and Evolution*, **9** (2018), 2206–2215.

91. J. S. Walker, M. W. Jones, R. S. Laramee, *et al.*, Prying into the intimate secrets of animal lives; software beyond hardware for comprehensive annotation in 'Daily Diary' tags. *Movement Ecology*, **3** (2015), 29.

92. J. J. Valletta, C. Torney, M. Kings, *et al.*, Applications of machine learning in animal behaviour studies. *Animal Behaviour*, **124** (2017), 203–220.

93. Y. S. Resheff, S. Rotics, R. Harel, *et al.*, AcceleRater: a web application for supervised learning of behavioral modes from acceleration measurements. *Movement Ecology*, **2** (2014), 27.

94. A. Mathis, P. Mamidanna, K. M. Cury, *et al.*, DeepLabCut: markerless pose estimation of user-defined body parts with deep learning. *Nature Neuroscience*, **21** (2018), 1281–1289.

95. K. Wurtz, I. Camerlink, R. B. D'Eath, *et al.*, Recording behaviour of indoor-housed farm animals automatically using machine vision technology: a systematic review. *PLOS ONE*, **14** (2019), e0226669.

96. M. S. Dawkins, R. Cain and S. J. Roberts, Optical flow, flock behaviour and chicken welfare. *Animal Behaviour*, **84** (2012), 219–223.

97. T. Nath, A. Mathis, A. C. Chen, *et al.*, Using DeepLabCut for 3D markerless pose estimation across species and behaviors. *Nature Protocols*, **14** (2019), 2152–2176.

98. R. P. Wilson, L. Börger, M. D. Holton, *et al.*, Estimates for energy expenditure in free-living animals using acceleration proxies: a reappraisal. *Journal of Animal Ecology*, **89** (2019), 161–172.

99. E. L. C. Shepard, R. P. Wilson, F. Quintana, *et al.*, Identification of animal movement patterns using tri-axial accelerometry. *Endangered Species Research*, **10** (2010), 47–60.

100. E. Browning, M. Bolton, E. Owen, *et al.*, Predicting animal behaviour using deep learning: GPS data alone accurately predict diving in seabirds. *Methods in Ecology and Evolution*, **9** (2018), 681–692.

101. J. Krause, S. Krause, R. Arlinghaus, *et al.*, Reality mining of animal social systems. *Trends in Ecology and Evolution*, **28** (2013), 541–551.

102. I. D. Couzin, Collective cognition in animal groups. *Trends in Cognitive Sciences*, **13** (2009), 36–43.

103. D. Papageorgiou, C. Christensen, G. E. C. Gall, *et al.*, The multilevel society of a small-brained bird. *Current Biology*, **29** (2019), R1120–R1121.

104. I. Psorakis, B. Voelkl, C. J. Garroway, *et al.*, Inferring social structure from temporal data. *Behavioral Ecology and Sociobiology*, **64** (2015), 875–889.

105. G. Roberts, Why individual vigilance declines as group size increases. *Animal Behaviour*, **51** (1996), 1077–1086.

106. R. Watson and P. Yip, How many were there when it mattered?: estimating the sizes of crowds. *Significance*, **8** (2011), 104–107.

107. M. S. Norouzzadeh, A. Nguyen, M. Kosmala, *et al.*, Automatically identifying, counting, and describing wild animals in camera-trap images with deep learning. *Proceedings of the National Academy of Sciences USA*, **115** (2018), E5716–E5725.

108. J. Reiczigel, Z. Lang, L. Rózsa and B. Tóthmérész, Measures of sociality: two different views of group size. *Animal Behaviour*, **75** (2008), 715–721.

109. T. W. Bodey, I. R. Cleasby, F. Bell, *et al.*, A phylogenetically controlled meta-analysis of biologging device effects on birds: deleterious effects and a call for more standardized reporting of study data. *Methods in Ecology and Evolution*, **9** (2018), 946–955.

110. N. Burley, G. Krantzberg and P. Radman, Influence of colour-banding on the conspecific preferences of zebra finches. *Animal Behaviour*, **30** (1982), 444–455.

111. D. Wang, W. Forstmeier, M. Ihle, *et al.*, Irreproducible text-book "knowledge": the effects of color bands on zebra finch fitness. *Evolution*, **72** (2018), 961–976.

112. P. P. G. Bateson, Testing an observer's ability to identify individual animals. *Animal Behaviour*, **25** (1977), 247–248.

113. C. L. Witham, Automated face recognition of rhesus macaques. *Journal of Neuroscience Methods*, **300** (2018), 157–165.

114. A. C. Ferreira, L. R. Silva, F. Renna, *et al.*, Deep learning-based methods for individual recognition in small birds. *Methods in Ecology and Evolution*, **11** (2020), 1072–1085.

115. T. Wey, D. T. Blumstein, W. Shen and F. Jordán, Social network analysis of animal behaviour: a promising tool for the study of sociality. *Animal Behaviour*, **75** (2008), 333–344.

116. D. R. Farine and H. Whitehead, Constructing, conducting and interpreting animal social network analysis. *Journal of Animal Ecology*, **84** (2015), 1144–1163.

117. G. H. Davis, M. C. Crofoot and D. R. Farine, Estimating the robustness and uncertainty of animal social networks using different observational methods. *Animal Behaviour*, **141** (2018), 29–44.

118. W. J. E. Hoppitt and D. R. Farine, Association indices for quantifying social relationships: how to deal with missing observations of individuals or groups. *Animal Behaviour*, **136** (2018), 227–238.

119. H. Whitehead, *Analysing Animal Societies: Quantitative Methods for Vertebrate Social Analysis.* (Chicago, IL: University of Chicago Press, 2008).

120. A. Sánchez-Tójar, J. Schroeder and D. R. Farine, A practical guide for inferring reliable dominance hierarchies and estimating their uncertainty. *Journal of Animal Ecology*, **87** (2018), 594–608.

121. H. de Vries, J. M. G. Stevens and H. Vervaecke, Measuring and testing the steepness of dominance hierarchies. *Animal Behaviour*, **71** (2006), 585–592.

122. T. Bedford, C. J. Oliver, C. Andrews, *et al.*, Effects of early-life adversity and sex on dominance in European starlings, *Sturnus vulgaris. Animal Behaviour*, **128** (2017), 51–60.

123. D. Chicco and G. Jurman, The advantages of the Matthews correlation coefficient (MCC) over F_1 score and accuracy in binary classification evaluation. *BMC Genomics*, **21** (2020), 1–13.

124. P. Ranganathan, C. S. Pramesh and R. Aggarwal, Common pitfalls in statistical analysis: measures of agreement. *Perspectives in Clinical Research*, **8** (2017), 187–191.

125. K. O. McGraw and S. P. Wong, Forming inferences about some intraclass correlation coefficients. *Psychological Methods*, **1** (1996), 30–46.

126. P. E. Shrout and J. L. Fleiss, Intraclass correlations: uses in assessing rater reliability. *Psychological Bulletin*, **86** (1979), 420–428.

127. T. K. Koo and M. Y. Li, A guideline of selecting and reporting intraclass correlation coefficients for reliability research. *Journal of Chiropractic Medicine*, **15** (2016), 155–163.

128. M. Tavakol and R. Dennick, Making sense of Cronbach's alpha. *International Journal of Medical Education*, **2** (2011), 53–55.

129. S. Mandillo, V. Tucci, S. M. Hölter, *et al.*, Reliability, robustness, and reproducibility in mouse behavioral phenotyping: a cross-laboratory study. *Physiological Genomics*, **34** (2008), 243–255.

130. M. J. Crawley, *Statistics: An Introduction Using R*, 2nd edn. (Chichester, UK: John Wiley & Sons, 2015).

131. A. Field, J. Miles and Z. Field, *Discovering Statistics Using R.* (London: SAGE Publications, 2012).

132. A. Field, *Discovering Statistics Using IBM SPSS Statistics*, 5th edn. (Los Angeles, CA: SAGE Publications, 2017).

133. S. McKillup, *Statistics Explained: An Introductory Guide for Life Scientists*, 2nd edn. (Cambridge, UK: Cambridge University Press, 2012).

134. R. T. Warne, *Statistics for the Social Sciences: A General Linear Model Approach.* (Cambridge, UK: Cambridge University Press, 2018).

135. H. Wickham, Tidy data. *Journal of Statistical Software*, **59** (2014), 1–23.

136. F. J. Anscombe, Graphs in statistical analysis. *Key Topics in Surgical Research and Methodology*, **27** (1973), 17–21.

137. A. Grafen and R. Hails, *Modern Statistics for the Life Sciences.* (Oxford, UK: Oxford University Press, 2002).

138. J. C. Douma and J. T. Weedon, Analysing continuous proportions in ecology and evolution: a practical introduction to beta and Dirichlet regression. *Methods in Ecology and Evolution*, **10** (2019), 1412–1430.

139. A. J. Wilson, D. Réale, M. N. Clements, *et al.*, An ecologist's guide to the animal model. *Journal of Animal Ecology*, **79** (2010), 13–26.

140. L. Z. Garamszegi, ed., *Modern Phylogenetic Comparative Methods and Their Application in Evolutionary Biology.* (Berlin/Heidelberg: Springer-Verlag, 2014).

141. A. J. Dobson and A. G. Barnett, *An Introduction to Generalized Linear Models*, 4th edn. (Boca Raton, FL: CRC Press, 2018).

142. B. M. Bolker, M. E. Brooks, C. J. Clark, *et al.*, Generalized linear mixed models: a practical guide for ecology and evolution. *Trends in Ecology & Evolution*, **24** (2009), 127–135.

143. S. Siegel and N. J. Castellan, *Nonparametric Statistics for the Behavioral Sciences.* (New York, NY: McGraw-Hill, 1988).

144. M. R. E. Symonds and A. Moussalli, A brief guide to model selection, multimodel inference and model averaging in behavioural ecology using Akaike's information criterion. *Behavioral Ecology and Sociobiology*, **65** (2010), 13–21.

145. M. Del Giudice and S. W. Gangestad, A traveler's guide to the multiverse: promises, pitfalls, and a framework for the evaluation of analytic decisions. *Advances in Methods and Practices in Psychological Science*, **4** (2021), 1–15.

146. R. McElreath, *Statistical Rethinking: A Bayesian Course with Examples in R and Stan.* (Boca Raton, FL: CRC Press, 2016).

147. T. R. Makin and J. J. O. de Xivry, Ten common statistical mistakes to watch out for when writing or reviewing a manuscript. *eLife*, **8** (2019), e48175.

148. A. Reinhart, *Statistics Done Wrong: The Woefully Complete Guide.* (San Francisco, CA: No Starch Press, 2015).

149. A. Gelman and H. Stern, The difference between "significant" and "not significant" is not itself statistically significant. *The American Statistician*, **60** (2006), 328–331.

150. M. Bateson, A. Aviv, L. Bendix, *et al.*, Smoking does not accelerate leucocyte telomere attrition: a meta-analysis of 18 longitudinal cohorts. *Royal Society Open Science*, **6** (2019), 190420.

151. J. M. Rohrer, Thinking clearly about correlations and causation: graphical causal models for observational data. *Advances in Methods and Practices in Psychological Science*, **1** (2018), 27–42.

152. J. Pearl, M. Glymour and N. P. Jewell, *Causal Inference in Statistics: A Primer.* (Chichester, UK: John Wiley & Sons, 2016).

153. The British Psychological Society, *Style Guide for Authors and Editors.* (Leicester, UK: British Psychological Society, 2018).

154. N. Percie, V. Hurst, A. Ahluwalia, *et al.*, The ARRIVE guidelines 2.0: updated guidelines for reporting animal research. *PLOS Biology*, **18** (2020), e3000410.

155. T. A. Lang and D. G. Altman, Basic statistical reporting for articles published in biomedical journals: the "Statistical analyses and methods in the published literature" or the SAMPL guidelines. *International Journal of Nursing Studies*, **52** (2015), 5–9.

156. K. Healy, *Data Visualization. A Practical Introduction.* (Princeton, NJ/Oxford, UK: Princeton University Press, 2019).

157. International Committee of Medical Journal Editors, *Recommendations for the Conduct, Reporting, Editing, and Publication of Scholarly Work in Medical Journals.* (ICMJE, 2019). www.icmje.org/icmje-recommendations.pdf

Index

Printed in the United States
by Baker & Taylor Publisher Services